Domain-Level Reasoning
for Spoken Dialogue Systems

Dirk Bühler • Wolfgang Minker

Domain-Level Reasoning for Spoken Dialogue Systems

 Springer

Dirk Bühler
Institute of Information Technology
University of Ulm
Albert-Einstein-Allee 43
89081 Ulm
Germany
dirk.buehler@uni-ulm.de

Wolfgang Minker
Institute of Information Technology
University of Ulm
Albert-Einstein-Allee 43
89081 Ulm
Germany
wolfgang.minker@uni-ulm.de

ISBN 978-1-4899-9148-5   ISBN 978-1-4419-9728-9 (eBook)
DOI 10.1007/978-1-4419-9728-9
Springer New York Dordrecht Heidelberg London

Springer is part of Springer Science+Business Media (www.springer.com)

Preface

Many existing natural language and spoken language dialogue systems are either very limited in the scope of domain functionality or require a rather cumbersome interaction. With an increasing number of application domains, ranging from unified messaging to trip planning and appointment scheduling, it seems to be obvious that the current interfaces need to be rendered more efficient.

The possibility to construct and to manage complex tasks and interdependencies with these applications requires a high cognitive burden from the user, which may be dangerous in certain environments (e.g., the well-known driver distraction problem in cars). Rather than preventing the use of such applications, however, it seems necessary to relieve the user as much as possible from the need to manage the complexities involved alone in his mind.

We argue that the system should serve as an integrated assistant to the user, i.e., it should be able to collaborate with the user to find a solution that fits the user's requirements and constraints and that is consistent with the system's world knowledge. In particular, the system's world model should include knowledge about dependencies between certain domains: In a travelling sales person scenario, for instance, the system could automatically calculate route durations and check for parking space depending on the user's calendar.

In this book we describe a logic-based reasoning component for spoken language dialogue systems. This component, called *Problem Assistant* is responsible for processing constraints on a possible solution obtained from various sources, namely the user's and the system's domain-specific information. The core processing is finite model generation. This inference technique attempts to find solutions that fit both the user's constraints and that are consistent with the Problem Assistant's rule base. Since the assistant interactively generates transparent information about its inference process, our approach provides the basis for incremental explanation dialogues and collaborative conflict resolution.

We also present findings on the implementation of a dialogue management interface to the Problem Assistant. The dialogue system supports simple mixed-initiative planning interactions in an application domain including logistics, evacuation and emergency planning. Although limited in terms of the number of entities modelled, this is still a relatively complex domain involving a number of logical constraints and relations. In our view, these logical constraints and relations form the basis for the collaborative problem solving behaviour that drives the dialogue.

Contents

List of Figures

List of Tables

1

Introduction

In view of recent advances of technology, such as the miniaturisation of increasingly powerful computing devices with a parallel reduction of costs, as well as the increased performance of specialised algorithmic solutions such as speech recognition, the prospect of a daily use of speech dialogue technology in a variety of situations becomes more realistic than ever [1]. In the automotive infotainment environment, for instance, product-level progress has been made that enables an interaction design to move away from spelling-based entry of parts of navigation destinations to a natural input of complete addresses. At the same time the vocabulary size has been increased.

Nevertheless, the car is also an example of an environment in which the transition from a command-and-control style of interaction using commands that result in immediate effects to a form of interaction that allows the user and system to collaborate on a deeper level has just begun. By "deeper level" we mean that the user has more complex goals and requirements than what is manifested at the surface level of a destination input, for instance. In fact, such an input is typically connected to a real-world activity of going to work or doing shopping. If dialogue systems and interaction technology is to leave the isolated environments and adopt a companion-like usage scenario then these kinds of activities and their structure should be modelled in order to make them an area where the system can provide assistance.

However, this will also incur a number of interesting challenges, some of which we address in this work. For illustration, consider the following interaction between a user U and system S: Assume U is on his way to a meeting.

```
U-1 > Add an appointment with Smith in Munich.
S-2 > Before or after your meeting in Stuttgart?
U-3 > How much time would we have if we meet before?
S-4 > The duration of the meeting would be at most 25 minutes.
U-5 > I need at least one hour.
S-6 > So, the meeting is after the one in Stuttgart?
U-7 > No, let's postpone the meeting in Stuttgart.
```

In utterance U-1, the user states a requirement to be added to his current schedule. Note that the information provided in the requirement is partial since, for instance, the meeting time or exact location is missing. Nevertheless, in the scenario the system does not just ask for values to fill these open "slots," but instead is able to reason with the given information in combination with the existing schedule. It comes up with two different possible solutions and presents the choice to the user in S-2, which also reminds the user of the other appointment in Stuttgart and possible new interdependencies. Instead of directly responding to the system request, the user chooses to initiate a subdialogue in U-3. The subdialogue is special in that it accepts one of the alternatives as a hypothesis rather than committing to it. This is also stressed by the use of the conditional mode ("would have", "would be") on the surface level. In S-4, the system provides an answer on the basis of the additional information and remains in this hypothetical context. One may note that, somewhat behind the scenes, the system also performs reasoning that integrates knowledge from different domains, namely the user's schedule as well as a navigation knowledge source that estimates travel between the locations. However, the user leaves that context by stating another requirement in the following utterance U-5. On that basis, the system is able to infer that the situation has changed and given the requirements introduced so far, one alternative has been ruled out. Thus, the system tries to confirm that the other option is chosen in S-6. However, instead of confirming the systems inference, the user rejects it in U-7. Logically, this constitutes a conflict between the set of established requirements. However, the situation is resolved by the user in the same utterance by hinting at which requirement (the starting time of the meeting in Munich) to relax.

While the construction of a complete end-to-end system, especially one robustly using speech and flexible natural language expressivity as illustrated in the scenario, is beyond the scope of this work, we address some issues that from our point of view are central to the scenario. These issues concern the flexible use of requirements in the dialogue, such that, for instance, creating subdialogues based on hypotheses becomes possible, and the way that the requirements are processed by the system in order to infer more information, such as options or conflicts. Our approach described in Section 1.3 will consist of an architecture and key components to enable these functionalities.

In the following, we introduce the dialogue systems framework our work is based on before providing more details concerning the problems present in a conventional architecture and we clarify how we propose to address them.

1.1 Spoken-Language Dialogue Systems

This section provides a brief introduction to the area of Spoken-Language Dialogue Systems (SLDS) technology [2, 3]. In order to put the description of dialogue systems architecture in context, we summarise the relevant commonly accepted levels on which language is analysed in linguistics: *Phonetics* describes the creation and discrimination of speech sounds (*phones*) that humans are in principle capable of producing and perceiving in order to communicate in a natural language. This also concerns the physical properties of speech sounds, as represented, for instance, in the frequency spectrum of an audio signal. On that basis, *Phonology* is concerned with the question which sounds are distinguished in a particular natural language. These are called *phonemes*. Phonology also studies, for instance, the influence of contextual pronunciation of phonemes. *Morphology* deals with the internal structure of words, as constructed out of simpler meaningful units, i.e. morphemes, and the processes that relate the internal structure to the surface appearance of the word. An example of a morphological process in English is the deletion of the stem-final "e" as in the following example: love+ing → loving. *Syntax* is concerned with the mapping of words to parts of speech and the recursive structure of phrases in terms of these parts. Traditionally, tree structures are used to represent some of these relationships. A common task of syntax in languages like English is the definition of the conditions of *agreement* in person and number between subject and predicate phrases in a sentence, for instance. *Semantics* deals with the representation of the meaning of an utterance. In a basic formulation, the task of semantics can be described as assigning a *logical form* to a sentence, i.e. identifying the truth conditions of a sentence. One aspect here is to take advantage of syntactical relationships in order to provide a compositional construction of meaning representations. *Pragmatics* concerns the interpretation of sentences within the context of a discourse. Speech acts formalise different ways that the same surface level utterance can be used. For instance, the sentence "it is cold here" may be interpreted as an informative statement or as a proposal for action depending on the context. Related aspects are discussed in more detail in Chapter 2.

Figure 1.1 illustrates the typical architecture of an SLDS as well as the modifications thereon proposed in our work (which will be discussed in Section 1.3). From an idealised perspective, one may view the operation of an SLDS as a cycle of separate processing stages. In this cycle, user utterances are translated from a surface (signal) level into a symbolic level. On that basis, the system determines its overall reaction and then renders it back to the surface level for the user to receive it. Then, another iteration may take place.

To some extent, different processing stages within this cycle reflect the linguistic levels introduced before. After the audio input is captured by an analog/digital converter the digital signal can be submitted to certain acoustic preprocessing steps, such as noise reduction and echo cancellation. Afterwards, the audio signal is usually translated into a feature vector representation on a time frame basis. In a typical automatic speech recognition system, sequences of these feature vectors are then translated into word sequences. On that basis, natural language understanding components using syntactical and semantic patterns knowledge translate the word-based representation into a semantic one which is independent of the utterance modality. As part of the dialogue

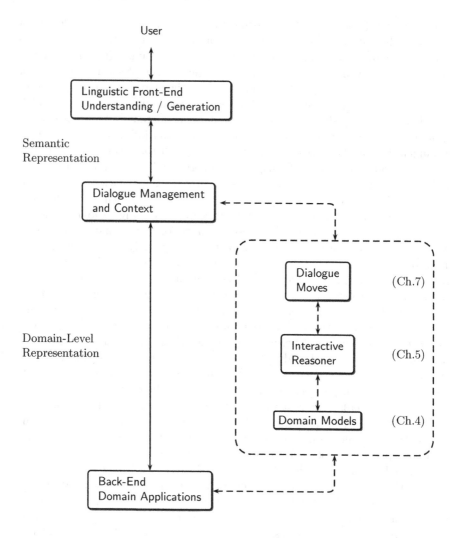

Fig. 1.1. Architecture of a Spoken Language Dialogue System (SLDS) and the extensions performed in the framework of this work.

management, an interpretation of this so far isolated utterance is performed in the context of the dialogue history (the previous processing cycles). For instance, the resolution of anaphoric references may be performed at this stage. Next, the system has to decide whether the interpretation of the utterance provided information relevant to the application level, i.e. information that can be used to, for instance, select an item from an application-dependent database. If so, the information is provided to the domain application and the results obtained determine the system's response. Otherwise, the system will try to use dialogue management strategies to obtain more information from the user. Whatever the content of the system response, it has to be translated back from the symbolic level to the surface. In our conceptual overview, this is done via a text generation module that produces the word-based response representation and text-to-speech synthesis which generates a corresponding audio signal. Upon receiving the system response, the user is expected to enter the next cycle by producing a new message to the system.

In terms of SLDS architecture, our focus in this work will be placed on the dialogue management and its interaction with the proposed additional components (cf. Figure 1.1). In doing so, we will assume a *linguistic front-end* comprising speech understanding and generation. We will describe the relevant basic technologies in more detail in the following.

Automatic Speech Recognition (ASR) is the process of determining the most likely word sequence W^* given an acoustic observation O and a probability model $P(W|O)$. Usually, the observation is a sequence of vectors of acoustic features that are calculated on a frame basis (typically around 10ms). Mathematically, this process can be formalised as:

$$W^* = \arg\max_W P(O|W)P(W)$$

This formulation shows the commonly accepted factorisation of the stochastic model into two distinct components. The *acoustic model* $P(O|W)$ assigns a probability to an acoustic observation given a word sequence. The *language model* $P(W)$ captures the probability of a word string solely based on the sequence of words. In many systems, the acoustic model is represented using Hidden Markov Models (HMM) [4]. They are usually based on phonemes with a dictionary encoding the relation between phonemes and words. The calculation of the most likely state sequence is accomplished using the Viterbi algorithm.

Natural language understanding (NLU) deals with the interpretation of a word-based representation of an utterance in terms of a semantic representation [5]. Thus, it comprises syntactical and semantic processes. One of the central challenges in natural language understanding is the resolution of ambiguities. One may distinguish two kinds of ambiguities: Lexical ambiguity concerns uncertainty arising from a word with different lexical meanings, or different syntactic functions. Structural ambiguity, on the other hand, arises from different possibilities concerning the relations between the words. Consider the example "I saw the (man with the telescope)" in contrast to "I (saw (the man) with the telescope)". Many approaches use rules that determine the interpretation of a word sequence. Such rules may represent syntactic or semantic constraints, or both. Rule-based approaches are credited for allowing to specify expert knowledge in detailed and concise formulations. On the other hand, they have often been criticised for being fragile and inflexible, especially in the presence of speech recognition errors [6]. Statistical (data-driven) models, on the other hand, are usually more robust and efficient [7, 8, 9]. However, they impose the problem of collecting and preparing a realistic and extensive corpus of training data.

Concerning the output direction of the linguistic front-end, the components that translate a semantic representation or logical form into an utterance understandable by the user are frequently concerned with two conceptually different stages: First, a word-based representation is constructed by means of text generation techniques. Then, the text-based representation is rendered as a speech utterance. Text generation involves selecting the content and the wording of the utterance. Text synthesis is concerned with selecting the correct pronunciations for the word-based representations and the intonational and prosodic modelling of the utterance. For spoken language dialogue systems, text-to-speech synthesis (TTS) is commonly applied in order to produce speech output based on a textual representation generated in the utterance generation phase. An alternative, using pre-recorded speech (or chunks thereof), is nowadays only used in restricted application settings where the variation of utterances is limited or the resources required for performing a TTS are not available.

1.2 Problem Setting

Isolated application domains. One reason why SLDS may not yet have used their full potential despite the noticeable increase in computational power is that each of the applications implemented in the SLDS approach has its own private and isolated knowledge state. The applications do not care about the user's "overall plan" (i.e. the general tasks that the user is trying to accomplish), or about interactions between the domain-specific knowledge states because the data of one application does not have a meaning for another. For instance, a meeting scheduled in a calendar application usually should not overlap with a travel itinerary.

Thus, redundancies and, even worse, logical inconsistencies between the knowledge states may arise, and the user is forced to cumbersomely manage the consistency on his own. This is not only time-consuming and error-prone, but may also result in mental overload or distraction. This may even be dangerous, for instance in mobile environments [10]. In addition, the inconsistencies may not be detected until all or some options for resolving them have already passed. In such a setting, the user has to decide which applications he needs to interact with and in which order. This becomes essentially impractical once an overall plan needs to be revised because of unexpected new information or changed conditions (for instance, a traffic jam delaying a travel).

Consistency reasoning would also be beneficial in applications where different yet overlapping aspects of a global knowledge state (overall plan) are encoded in different representations or data formats. For instance, information about a business travel may be spread between (and redundantly represented in) the traveller's calendar and different electronic administrative documents, such as travel applications or travel expense reports. In such scenarios, one would like to avoid inconsistencies between the different data, or resolve them if they are present, and to propagate changes occurring in one representation to the other representations.

Lack of transparency. One of the key problems of many contemporary computer interfaces is the opaqueness of their processing, and the lack of an appropriate explanation facility. Conventional software often behaves as a black box, i.e. it does not explain its internal reasoning to a satisfactory extent. There is not a sufficient amount of transparency concerning what the computer is doing, is about to do, or has already done. As a consequence, conflicts and logical inconsistencies in the data cannot be traced back effectively to their origins, if detected at all. And, as a further consequence, they cannot be resolved. Consider for instance, the task of moving a folder from one partition to another which does not have enough free space. Typically, the system will try to do it and, at best, recover in a way that keeps the original states intact. Nevertheless, a better approach would have been to predict the problem beforehand because the prerequisites of the operation are not met. One of the reasons is that conventional software applications implement their logic in a very rigid and inflexible way, for instance, they allow only very specific use cases of this logic and only very specific flows of information are actually executable in the software. Moreover, in many cases information concerning "what went wrong" is only expressed in an insufficient way (such as error codes) or is lost altogether.

Another problem concerns the need for hypothetical reasoning. Conventional software deals with only one knowledge state at a time, for instance, one schedule of meetings. This makes it difficult to compare different alternative possibilities, or *scenarios*. Hypothetical reasoning is also a prerequisite to be able to predict the consequences of planned activities rather than execute the actions and observe the consequences. The consequences may be irreversible, so as much as possible has to be known in advance. Instead, today's applications often execute actions that change their knowledge state and perhaps have some effects that become obvious only after the actions have been executed. However, there is no declarative reasoning in advance. In particular, even if the user is asked if he wants to execute a certain action, it is frequently the case that it not clearly stated why an action is necessary, or whether it is just a suggestion among other alternatives, or what the consequences of the action will be. Alternative scenarios can also arise from underspecification in dialogue, for instance, when the user provides incomplete or vague information. In a trip planning domain, for instance, the user may want to plan the details of alternative solutions and make a decision later based on a comparison of the alternative parameters. One of the few SLDS that provides hypothetical reasoning is, for instance, TRIPS [11]. However, we argue that the way how this is accomplished could be improved (cf. Section 2.8.2).

1.3 Proposed Solution – A Logic-Based Framework

Our proposed solution is based on the general architecture of SLDS, cf. Figure 1.1. We argue that the problems identified in the previous section can be addressed by focusing on the interface between the dialogue management and the domain level, and the related proposals will be presented in the following section.

In order to address the problems laid out before, our work aims at achieving the following goals: Domain application knowledge will be integrated in a shared knowledge state, rather than isolated from each other. A common reasoning service will provide inferences concerning interactions between applications. In particular, it is to provide transparent and comprehensive access to its reasoning. To this end, it will include an interactive protocol to guarantee that the reasoning processes can be managed in an interactive fashion. A dialogue management approach will make the reasoning useful for a better interactive user experience. The dialogue management will be based on domain-independent dialogue functions which take advantage of the interactive reasoning engine. The next paragraphs provide more details on these aspects.

Dialogue interaction. From a dialogue systems perspective the main goal of this work is to provide a novel concept of dialogue interaction which addresses the problems described in the previous section. Some of the key concepts of our dialogue management approach are illustrated in Figure 1.2.

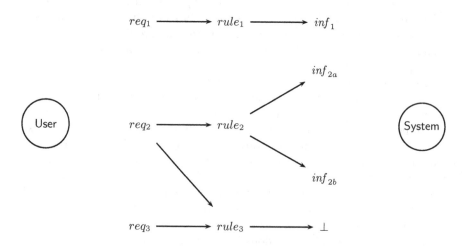

Fig. 1.2. The proposed requirements-inferences dialogue management approach.

We denote by a (user) *requirement* domain-level information introduced by the user to describe a domain task in terms of goals, constraints, conditions, or assumptions. We usually avoid these other terms in order to prevent confusion. Typically, user requirements are atomic expressions that state relations between objects on the domain level. A (system) *inference* is any information that the system derives on the basis of the provided requirements by performing logic-based reasoning. Here, the inf_i refer to atomic expressions. Three different kinds of inferences may occur: First, the inference may result in a single atom (e.g. inf_1, not identical to $\bot$). In that case, the system has concluded some information on the basis of the requirements which may or may not need to be discussed with the user. Second, if the inference is $\bot$, a *conflict* has been detected. This means that the requirements are inconsistent, i.e. they cannot be fulfilled according to the system's knowledge. For instance, in Figure 1.2, the requirements req_2 and req_3 are in conflict. In this case, the user has to revise his requirements (which will be discussed shortly). Finally, if the inference results in *alternatives* (for instance, inf_{2a} and inf_{2b}) the requirements present an underspecified situation that needs to be resolved. Apparently, the inferences map to a logic-based reasoning based on preconditions and postconditions. In the figure, the preconditions are depicted as incoming arcs to a rule node, whereas the postconditions are outgoing arcs. Importantly, preconditions are interpreted conjunctively while postconditions are disjunctions. In fact, this kind of reasoning is commonly referred to as *forward inference* in logic. The inference process can be iterated by the system, such that a lattice of inferences may be constructed.

The goal of the dialogue interaction between the user and the system is to collaboratively construct a *solution*, which may be defined as a set of requirements and inferences such that (i) no conflict is implied, and (ii) all appearing disjunctions are resolved. As we will see this corresponds to the logical concept of a *model* of the requirements (cf. Chapter 3). We refer to any subset of a model, as a *partial solution*. In order to disambiguate alternative postconditions, such as inf_{2a} and inf_{2b}, the system may propose one of the inferences to the user. In that case, we refer to this inference as a *suggestion*. Qu et al. [12, 13] have proposed a similar approach in the framework of constraint satisfaction. However, they mainly focus on overconstrained situations.

We argue that this proposed model is suitable for usage in practical dialogue systems [14] in many interesting application domains. In particular, using this proposed model many interesting dialogue phenomena such as disambiguation, conflict resolution, and hypothetical reasoning may be expressed in a way that is suitable for system implementation (and with a relatively small number of concepts).

Interactive reasoning. The problem of transparent reasoning and inference is addressed by providing a shared reasoning and proof management service to all domain applications on the basis of the common modelling. This Common Interactive Domain-level Reasoning Engine, called CIDRE, manages the information provided by different sources and controls the generation of inferences based on this information. It also maintains the inferences in a proof database. This enables a user or application to trace back any inference the reasoner has produced to its respective underlying assumptions. This way, a basis for an effective explanation facility and thus, for interactive conflict resolution is provided.

The implementation of the CIDRE reasoning engine is based on existing computational procedures for first-order model generation. The result of a reasoning process is the assurance that a set of requirements can be fulfilled given the current state of knowledge (plus perhaps additional assumptions to be negotiated with the user). Alternatively, the set of conditions is inconsistent (and the proof of its inconsistency is included). If the specified conditions are satisfiable, the result of the reasoning includes information concerning whether more than one solution exists and which decisions distinguish them from each other. Partial results may already contain important inferences, for instance showing dependencies between application domains (i.e. a scheduled meeting may create the need for trip planning and also put constraints on the travel itinerary), as well as information about decisions to be made.

In contrast to conventional reasoning engines, such as many automatic theorem provers, CIDRE has been designed from the beginning as a component that interacts with external information sources, such as the user. To this end, the reasoner provides an interactive protocol that allows an incremental and interactive access to concurrent reasoning processes. This is achieved by an asynchronous message protocol that enables an exchange of information during the reasoning process, not just at the end. Thus, the reasoner is able to communicate inferences, decisions, and assumptions at the time when they are relevant and the user may react to this information without delay. As a consequence, the search space for a complete solution can be delimited, because the reasoning process will incorporate the new information. When applied to a logistics domain, for instance, the reasoner may come up with the question what means of transportation to choose, even if it has not yet completed a solution for any of the options. It also may incrementally rule out choices as soon as it detects that assuming this choice would lead to a conflict with other conditions. To the best of our knowledge our approach is the only one of its kind that combines a model generation procedure with an interactive protocol that allows comprehensive access to the reasoner's inferences as well as controlling the reasoning engine.

Shared representation. Our proposed solution is based on a logic framework. It addresses these problems by integrating the knowledge states of different domain applications and synchronising the representations whenever new information becomes available. The integration results in a global view of the available information. Applications may only be interested in certain aspects or projections of the global view. The framework's knowledge representation is domain-independent. It adds a common semantic layer that interconnects application domains. On this layer each application's data has a common, defined semantics, even though each application may also manage private extensions of the data that make sense only for that application. For instance, a business meeting planning application may be interested in constraints such as available devices in a meeting room whereas a travel planning application would only be interested in the fact that the user is planning to be in a certain location at some definite time. The global view of the information represented ensures that conflicts between domain applications can be detected and that new information, e.g. due to the resolution of a conflict, will be available to all applications.

We argue that common modelling, such as a library of shared concepts or a classification hierarchy, is required. These concepts may be quite abstract, i.e. representing notions like physical object, and can be further refined into more detailed domain-specific concepts. Providing common concepts is the basis for the integration of different application domains. In this work, such a base library and specialisations for sample domains have been developed. In addition, we describe a way of modelling dynamic properties of entities. Using this approach, different phenomena such as changing positions when travelling, or changing cargo states in a logistics domain can be modelled generically. This formalisation has been developed in particular to be usable with the interactive reasoning engine introduced in the previous paragraph.

Architecture. We essentially view interactivity as the key to providing the enhanced reasoning capabilities to the user. Therefore, a corner stone of our work is the development of an architecture that makes use of our reasoning engine CIDRE. In such an architecture, the dialogue manager is the central processing component. Using CIDRE, novel and powerful dialogue functions can be implemented in an elegant and efficient way in the dialogue manager. Moreover, care has been taken to keep these functions domain-independent, such that portability to new domains and applications is ensured.

With regard to the design of the architecture two substantially different approaches have been investigated. The first was to use a dialogue scripting language (VoiceXML) and provide an interface to the reasoning engine. The aim of the research was to determine whether such form-based frameworks can be extended to implement our requirements-inferences dialogue management approach. This has resulted in a prototype in a restaurant selection domain.

To address some of the issues discovered in the form-based approach, a second prototype has been developed including a basic modular system architecture. In this prototype, the dialogue functions have been implemented in an information state-based approach [15], providing a novel set of dialogue moves as the elementary actions. In addition, the architecture prototype realises tighter and more efficient integration of the processing modules.

1.4 Document Structure

This book is organised as follows: In Chapter 2, different approaches to dialogue modelling and human-machine interaction are reviewed. We illustrate how the pioneering work on modelling communicative acts led to the development of various plan-based approaches. For our work, the most important result is a model that separates the domain (or application domain) level from the discourse and problem-solving levels. The domain level is where our approach to modelling application knowledge and our proposed reasoning engine operate. However, our work will not lead to a plan-based model of dialogue. Instead, our approach to dialogue management is oriented at two other technological trends that will also be discussed. These include dialogue scripting languages like VoiceXML and information state-based approaches. In addition, implemented systems related to our aims are reviewed and we discuss how our approach can complement this research.

Logic-based approaches, in particular first-order ones, are the pre-dominant approach to modelling domain knowledge. Therefore, we will provide a brief introduction to the main concepts of First-Order Logic in Chapter 3. Apart from these concepts and notations, this chapter will also describe two important aspects of the state of the art that are relevant to our work. The first is a formalisation of actions and events in a temporal logic. An adaptation of this theory will be an essential part of our approach to domain modelling. The other aspect is the introduction of an inference technique called first-order model generation, which will be the basis of our proposed reasoning engine.

On that basis, Chapter 4 will be dedicated to our approach to logic-based domain modelling. As mentioned in the previous paragraph, it includes an approach to formalising events, which has been formalised with concern for making it a feasible tool in combination with model generation. In addition, this chapter also includes the description of a modular library of domain theories. We have developed this library on the basis of the event theory, and it constitutes a corner stone in our work toward the integration of different application domains. In Chapter 5 we describe the design and implementation of our reasoning engine CIDRE which operates on the basis of the domain modelling outlined before. We also discuss its application to sample problems. One main innovation consists in the possibility to use this engine as an interactive component, in particular, within a dialogue systems architecture.

In Chapter 6 we describe a first prototype that we have developed to integrate the reasoning engine into a dialogue systems framework. This framework is based on VoiceXML and includes a compilation approach that enables us to tightly integrate code accessing the reasoner's interactive protocol. Abstracting from the prototype we discuss the potential of using VoiceXML as a basis for further applications, as well as the problems we encountered. We have addressed these problems by the development of a more generic dialogue management approach, which will be presented in Chapter 7. Instead of relying on VoiceXML, this approach is cast in terms of information states and dialogue moves. We describe how novel dialogue moves can be realised using the interactive protocol provided by CIDRE. This approach to dialogue management also led to a revised architecture which will be presented together with a (second) prototype implementing it in Chapter 8.

Finally, in Chapter 9 we provide a summary our work as well as an outlook to possible future research directions. This concerns both the application of our work in various areas of current research and possible extensions of our modelling and reasoning approach.

2

Fundamentals of Dialogue Systems

2.1 Introduction

This chapter presents an introduction to the fundamental concepts and developments that shaped the field of dialogue systems research, and thus presents the basis of our work. Speech act theory, together with the plan-based notion of rational agency, has been the predominant framework of most theoretical approaches to both human-human and human-computer dialogue and is still highly influential today. However, conversely to this process that may be called "top down," as it starts at the high level of intentions, in the dialogue systems community a "bottom up" movement has become popular. It can be characterised as aiming to build actual systems, especially using speech, that can realise specific dialogues at the surface level without a major theoretical or computational overhead. This has led to the development of dialogue scripting approaches, such as VoiceXML, among others.

We see our work as a building block in a compromise between the two directions. Although for our approach we do not adopt a plan-based framework which in variation is the basis of much of the research described this chapter, we recognise that the plan-based approach has led to important notions and modelling approaches, such as the separation of the domain level from other levels of representation in the discourse. Our approach to dialogue management, instead, is based on two later developments: VoiceXML and information state-based approaches.

The chapter is organised as follows: In the following we will give a brief overview of the development of the plan-based approach to analysing discourses and dialogues, before discussing some of the developments in more detail in Sections 2.2, 2.3, and 2.4. Sections 2.5 and 2.6 are dedicated to the converse trends of dialogue scripting languages and information state-based approaches. Finally, we will discuss a selection of implemented systems that exhibit the features relevant to our work in Section 2.8. On this basis, we will draw conclusions as to our architecture and design decisions.

Trying to understand and formalise how humans use language in order to achieve goals has a long tradition in linguistics and philosophy. Rather than isolated utterances most analyses address *discourses*, i.e. ordered sequences of utterances (or conversational acts) that aim to achieve a common purpose in the audience. The goal of modelling discourse is to understand how the individual parts of the discourse contribute to the overall interpretation of the discourse. A discourse can be written or oral, and it can be a monologue, or a dialogue between two or more discourse participants. Although a body of letters or e-mail messages can constitute a discourse, most attention is usually paid to face-to-face discourses.

A *dialogue* can be regarded as a discourse that involves more than one active participant, in the sense that each participant contributes to the discourse. A dialogue is an exchange of messages (utterances) between these *dialogue participants*. In a dialogue, a sequence of consecutive utterances by one dialogue participant is grouped into a *turn*. *Turn taking* refers to one participant assuming or yielding the turn in a dialogue.

Much of today's theoretical understanding of conversational acts is based on the works of Austin [1962] and Searle [1969]. They introduced the notion of a *speech act* which recognises the fact that speaking is a form of *acting* in the sense that the speaker wants to achieve some effect in the mental state of the hearer. Examples of speech acts include asserting, questioning, and commanding.

Grice [1975] introduced the notion of *conversational maxims* as a realisation of a cooperativity principle in dialogue . These maxims constitute rules that are employed commonly (and sometimes without conscious thought) in order to interpret utterances. The basic idea is that using the rules the hearer of a message can draw inferences (*conversational implicatures*) about the speaker's intentions behind an utterance. These inferences allow the hearer to further interpret an utterance beyond the literal meaning of the words spoken. Grice distinguishes between the maxims of quantity, quality, relevance, and manner. For instance, the maxim of relevance would allow the hearer to infer that the speaker considers the information "it is cold here" to be actually relevant in the discourse and that he wants to achieve some effect in the hearer by uttering it.

Building upon Searle's work plan-based models have been developed that view speech acts as actions within a more general plan representing the intentions of the speaker [19, 20]. Formalising speech acts in a plan-based framework allows to apply and adapt the achievements in general planning research to the problem of discourse analysis. The framework also furthers the formalisation by requiring applicability conditions (preconditions) and effects (postconditions) to be specified for speech acts, which may then be regarded as *planning operators*. A further advantage of the plan-based framework is that it may be the basis for understanding how conversational actions are intertwined with physical non-communicative actions. Cohen and Perrault concentrate on the formalisation of REQUEST and INFORM acts. In doing so, questions can be reduced to requesting to be informed. They formalise the planning operators in a modal logic framework with a possible worlds semantics. Cohen and Perrault present a language for describing planning operators and states of the world. They do not focus on how the speech acts can be realised using words or how a hearer can recognise speech acts from an utterance.

Grosz and Sidner [1986] present a theory of discourse that introduces the distinction of three interrelated structures, namely the *linguistic structure*, the *intentional structure*, and the *attentional structure*. This distinction serves to separate the intentional structure from the other aspects of the discourse and thus allows to develop specific models for it. The distinction has since been adopted by a large body of subsequent work.

Refining the notion of plans as complex mental attitudes [22], Grosz and Kraus [1993, 1996] develop the notion of a *SharedPlan*. A *SharedPlan* is a plan developed by a group of agents in order to achieve a shared goal. The goal is to [24, p.1]

> [...] provide a basis for constructing computer agents that are fully collaborative as well as to provide a framework for modelling the intentional component of dialogue.

SharedPlans aim to explain how group activities can be decomposed to the level of individual plans and actions. The model defines how agents can identify ways to perform a group action (i.e. find a *recipe* and identify the necessary parameter values) and how to avoid performing conflicting actions. Lambert [1991] introduces a tripartite model of discourse that emphasises the distinction between the communicative level, the problem solving level and the domain level. This distinction is relevant to our work. Essentially, our proposed reasoning engine applies to the domain level, while processes on the problem solving and discourse levels are coordinated by the dialogue manager. We will discuss some of the developments in more detail in the next sections, before moving on to describe actual system implementations.

2.2 The Structure of Discourse

This section reviews fundamental work concerning how to model different aspects of discourse. Grosz and Sidner's theory of discourse [21] is based on the distinction of three interrelated structures. Each of these captures a different aspect of the utterances that constitute the discourse. The utterances may be in written or in oral form and the discourse may involve only one, two, or many conversational participants (CPs). The theory is generic in the sense that it does not presuppose a human-human or a human-computer discourse. It tries to be applicable to all kinds of discourses. Thus, the aim of the work is to provide insights for psychologists analysing human discourses, as well as researchers and engineers involved in constructing human-machine dialogue systems.

The three different constituents of the discourse model are described in the following: The *linguistic structure* contains the sequence and structural relations of the utterances of the discourse; it aims to represent the "natural aggregation" of utterances into *discourse segments*. The kinds of aspects that are dealt with in the linguistic structure are related to linguistic phenomena, such as anaphora resolution (referring expressions), question-answer pairs and the like. The aggregation of utterances into discourse segments is analogous to the aggregation of words in an utterance, with each component playing a specific role in the entire structure. Discourse segmentation is not solely based on the order of the utterances in the discourse. Non-consecutive utterances may be in the same segment. Certain psychological and psycholinguistic effects (such as differences in pauses lengths) provide evidence for discourse segments. In addition, cue phrases and prosodic variations may also act as markers and are indications of discourse segment boundaries. The authors argue that these markers can be distinguished as to whether they explicitly influence the intentional or the attentional structure. The structural relations between discourse segments reflect the relationships among components of the intentional structure. This structure is discussed in the following.

The *intentional structure* deals with the "discourse-relevant purposes" that are part of the discourse. These kinds of purposesand the structure of these purposes are related to the coherence of the dialogue and also to the question if a sequence of utterances constitutes one or many discourses. For instance, a discourse that serves the single goal of convincing an audience and in which this purpose is pursued in a structured manner is usually regarded as very coherent. Although a discourse participant may have more than one purpose in participating (for instance, to entertain, as well as to describe some event), one purpose of the discourse is seen as foundational and is called the *discourse purpose* (DP). This intention provides the reason for the speaker conveying the particular content of this discourse. At a finer level of detail, each of the discourse segments is associated with a *discourse segment purpose* (DSP) which describes how the discourse segment contributes to the overall goal of the discourse (the DP). Discourse purposes are intentions that are to be recognised by the audience (rather than being private). In fact, some of their effects are not achieved unless the intentions are recognised. Compliments, for instance, are intended to be recognised as such. The authors distinguish the following types of intentions that are candidates for serving as DPs or DSPs: These intentions express a "want" that some other agent

- perform a physical act,
- believe some fact, or believe that some fact supports another,
- intend to identify some object, or
- know some property of an object.

The two major structural relations that link discourse purposes in a hierarchical fashion are dominance and satisfaction precedence. A discourse segment purpose dominates another if the latter contributes to the first (i.e. the latter is part of a plan to achieve the first). Satisfaction precedence introduces a kind of temporal ordering that specifies that certain discourse segment purposes have to be satisfied before others.

The third element of the discourse model is the *attentional state*. It represents the discourse participants' focus of attention and contains information about the salient objects, their properties and relations. It is a dynamic structure that evolves in parallel with the other component structures. It may change more substantially while, for instance, the linguistic structure is growing in a monotonic fashion. The attentional state is regarded a property of the discourse, and not one of the discourse participants. The attentional state is organised as a set of so-called *focus spaces*, the focusing structure. These focus spaces can be added and deleted in an update process that is governed by so-called transition rules. The whole process is referred to as *focusing*. Each focus space includes the entities that are salient in a particular discourse segment, as well as the discourse segment purpose. Focusing associated the focus space with the discourse segment.

The different constituents contain the information necessary for the conversational participants to decide how to interpret a new utterance, i.e. how it fits into and modifies the given discourse structure. The authors note the importance to distinguish between the three constituent structures for explaining certain discourse phenomena. The authors also argue that their theory of discourse is not a theory of discourse meaning. They point out, however, that such a theory would most likely include a theory of discourse structure as a building block.

2.3 A Hierarchical Model of Intentions

This section reviews work concerning the distinction of different levels of the representation of actions and intentions in a dialogue. While the distinctions introduced in the previous section may be presented as a horizontal split, the model presented in this section focuses on the intentional structure and splits it vertically (cf. Figure 2.1). The relevance to our approach stems from the introduction of the notion of *domain level*.

In order to represent user intentions in a plan-based approach to dialogue modelling, Lambert [1991] introduces a distinction between three kinds of actions and intentions: *communicative or discourse actions, problem solving actions*, and *domain actions*. This corresponds to a tripartite hierarchy of levels (cf. Figure 2.1). At the lowest level, the *discourse* or *communicative level*, discourse acts are represented and disambiguated. Discourse acts create and modify goals and acts on the second, intermediate, level, the *problem solving level*. In turn, actions on this level modify the *domain level* (the highest level) where the actual goal of the planning process is specified. On the domain level, goals such as travelling by train are represented. The problem solving level contains actions such as instantiating a parameter in a plan to travel, and the discourse level handles goals such as obtaining information, clarification, or expressing uncertainty.

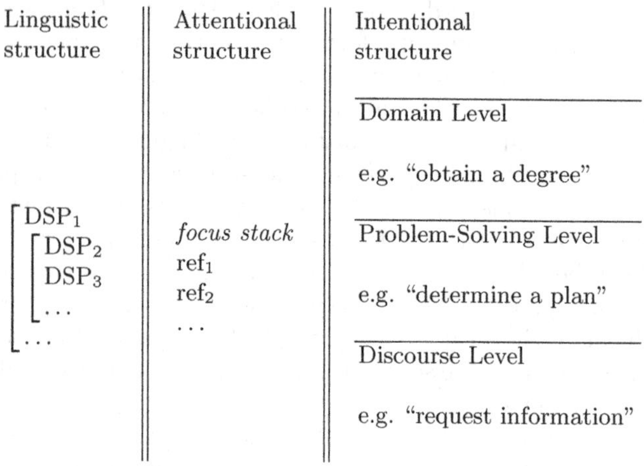

Linguistic structure	Attentional structure	Intentional structure
		Domain Level
		e.g. "obtain a degree"
$\begin{bmatrix} \text{DSP}_1 \\ \begin{bmatrix} \text{DSP}_2 \\ \text{DSP}_3 \\ \cdots \end{bmatrix} \\ \cdots \end{bmatrix}$	*focus stack* ref_1 ref_2 $\cdots$	Problem-Solving Level
		e.g. "determine a plan"
		Discourse Level
		e.g. "request information"

Fig. 2.1. The hierarchical structure of dialogue [25] in the context of the discourse structure proposed by [21].

Actions on the lower levels modify plans on the higher (more abstract) levels. This implies that, when executed, discourse actions can modify the current problem solving plan. For instance, a discourse-level "Surface-Inform" (see below) can define a parameter value in a problem solving-level plan. In turn, completed problem solving actions are used to modify the domain plan. For instance, choosing a particular problem-solving plan may introduce domain-level actions. Conversely, information needs related to modifications on the higher levels initiate actions on the lower levels. For instance, using a particular plan recipe on the problem-solving level may initiate a "Surface-Request" due to open parameter values in the recipe.

The three levels are represented as tree structures, making explicit the hierarchical ordering (and contribution relation) between the actions. The three levels are linked according to the modification/initiation relation described before. Within the tree structures, hierarchical plans ("subaction-arc") are constructed. The leaves of the discourse-level tree consist of surface realisations such as "Surface-Inform" or "Surface-Request". These are parts of a more high-level communicative plan. For instance, one may issue a "Surface-Inform" in order to make a subsequent "Surface-Request" acceptable (as required and defined by the discourse-level plan library.) "enable-arc" links connect the discourse level to the problem solving level (as well as the problem solving level to the domain level).

One important insight is how to distinguish between the domain level and the problem solving level: The authors note that the planning agent (e.g. the user) is the agent of all domain-level actions whereas at the problem-solving level, both participants collaborate as joint agents. Thus, the problem-solving level can be considered a *meta-level* with respect to the domain level. In later work the question of collaborative conflict detection and resolution on the basis of the tripartite structure is addressed [26, 27].

Ramshaw [1991] presents an alternative three-layered model of discourse. However, his focus is on using a middle layer between the domain layer and the discourse layer in order to allow for the exploration of domain plans. For instance, in a banking domain, the exploration of a domain plan that involves opening a banking account may consist of exploring different variants of that plan using different instantiations of parameters. Ramshaw argues that the different layers are necessary to capture different phenomena occurring when the speaker "is on a certain level" or when he switches levels.

2.4 Collaborative Negotiation

This section introduces the concept of collaborative negotiation which is relevant to our work in the sense that it proposes a methodology of how dialogue participants can cooperate to reach agreements concerning beliefs in dialogue.

The notion of collaborative negotiation refers to cooperative behaviour where two or more participants try to establish a common set of beliefs in order to solve a shared problem. In adversarial negotiation, on the other hand, one participant tries to win over another by achieving an agreement that is more advantageous to him than to the other participant. This is ruled out in collaborative interactions. In particular, the absence of deception or intentional misinformation is assumed in collaborative negotiation.

Sidner [1994] has presented an "Artificial Discourse Language for Collaborative Negotiation". This discourse language aims at modelling the interactions that occur in collaborative negotiation. The language consists of messages to propose, reject, counterpropose beliefs, or seek supporting information for certain beliefs. The term "belief" refers to arbitrary statements in an application domain. These messages are postulated in order to abstract from natural language such as English and to formalise part of the intentional structure in the sense of [21] (cf. Section 2.2). Two dialogue participants, called *agents* in the authors' terminology, send out messages of different types to each other. After each message is sent, certain logic conditions which are part of the language definition can be assumed to hold. This process is not necessarily monotonic, a belief revision system (or truth maintenance system) is assumed in order to drop certain previous assumptions which are inconsistent with new knowledge. This is necessary, because agents may have revise some of their beliefs. However, the details of that mechanism are not discussed. During the discourse a stack of "open" and "rejected" beliefs is maintained.

We summarise some of the most important message types here:

- ProposeForAccept: PFA $agent_1$ $belief$ $agent_2$: This message is sent by $agent_1$ to $agent_2$ in order to propose $belief$ as a mutual belief. $agent_2$ should acknowledge the message (which does not imply accepting the belief). The belief is added to the "open" set.
- AcknowledgeReceipt: AR $agent_1$ $belief$ $agent_2$: This message is sent by $agent_1$ to $agent_2$ in order to signal that $agent_1$ has received the proposal by $agent_1$. It does not imply that $agent_1$ is accepting the belief. The message is still important because $agent_2$ can now assume that both know as a mutual belief that $agent_2$ holds $belief$.
- Reject: RJ $agent_1$ $belief$ $agent_2$: This message is to signal that $agent_1$ does not hold $belief$. $belief$ is retracted from the "open" set.
- AcceptProposal: AP $agent_1$ $belief$ $agent_2$: Receiving this message, $agent_2$ can infer that both agents share $belief$. Furthermore, the proposed $belief$ is retracted from the "open" set.

The beliefs that are communicated are constructed from the operators BEL (belief), INT (intend), MB (mutual belief), and the following primitives: Communicated(), Should-do(), Achieve(), Supports(), Provide-Support(), Tell-if(), Identify(), and Able().

Questions and commands can be expressed using the introduced primitives: For instance, a question of the form "Why X?" is analysed as follows:

```
(PFA agent1 (Should-Do agent2
   (Provide-Support X context)) agent2)
```

Similar analyses can be assigned to utterances of the form "What is X?" (using Identify() instead of Provide-Support()), "Can you X?" (Able()), or factual questions like "Did John come?" (without specific primitive).

The authors also develop a process model that tries to explain how each message updates the dialogue state. Building on the concept of a stack-based attentional state introduced in [21], stacks for open and rejected beliefs are introduced. However, the author assumes a general form of "automated belief revision system to track all the mutual beliefs" in order to make the process model executable. Sidner's artificial language was also used as a basis for later information state-based approaches [30, 31].

2.5 Dialogue Scripts and Voice Browsers

In contrast to the traditional, highly generic, approaches previously outlined, this section presents a completely different approach to developing systems that enable a dialogue interaction, especially using spoken language. This section mainly focuses on VoiceXML, but this is by no means the only dialogue scripting language, nor is it the first one. However, it is arguably the most successful one as far as adoption by developers is concerned.

In the late 1990s, the World Wide Web (WWW) was becoming vastly popular and so did hypertext mark-up languages (HTML) for publishing content in that medium. Soon it was recognised that not only static content was to be published, but in addition interactive applications which in principle rely on the same kinds of user input that usual graphical computer applications need. Hyperlinking was not sufficient to allow for this kind of input, thus the idea of an HTML *form* was born, and – despite its limitations and different incompatible enhancements – has become vastly successful.

It was that development that inspired the growing telephony speech industry to establish a standard for web-based voice applications that should achieve a similar success. The idea was to develop an approach that would allow to quickly deploy basic voice-based applications without a constraining theoretical overhead. Thus, the Voice Extensible Mark-up Language (VoiceXML) 1.0 specification was proposed by a consortium of major telecommunications and software companies (AT&T, Motorola, Lucent, IBM, among others) in 2000. The current W3C recommendation version 2.1 adds certain extensions and improvements on this basis [32]. A simple VoiceXML application is illustrated in Figure 2.2.

```
<vxml>
  <form id="form1">
    <block>Hello user!</block>
    <field name="username">
      <grammar uri="names.srgs" />
      <prompt>What is your name?</prompt>
    </field>
  </form>
</vxml>
```

Fig. 2.2. A simple VoiceXML application.

The most important aspect of the VoiceXML specification is the fact that it defines an abstract processing model for VoiceXML applications, called the *Form Interpretation Algorithm* (FIA). The FIA assumes that initially the system has the turn, and it determines which actions the system should take, in particular how to query for user input, and how to respond to user input or failure conditions (events), if any. The basic structure that the FIA operates on is the *form*. It consists of *items* which belong to different classes, the most common two of which are *fields* and *blocks*. Similarly to forms in HTML, a VoiceXML form consists of smaller units that handle individual form components, i.e. these units allow the user to provide or obtain values of individual task parameters. In HTML, a common element is the <input> element, while in VoiceXML, the respective unit is the <field>. The VoiceXML approach to form-based dialogues is to iteratively update a dialogue state consisting of a record structure of simple named values. For instance, consider a record containing a departure and destination city, both represented as strings. We summarise the form processing in the following. In the context of the current form, the FIA goes through the following phases:

1. The *Select phase* determines the next form item by determining which items are have not yet obtained a value and are applicable (not prohibited by item guard conditions). If no such item can be determined, the dialogue ends.
2. The *Collect phase* is responsible for playing the item's prompt (or performing executable actions specified in a block). In addition, the speech recognition subsystem is configured to listen for user input and to analyse it with the adequate grammars activated.
3. The *Process phase* processes the value or event obtained from the recognition attempt in the Collect phase. If an exception event occurs because, for instance, the user did not speak or could not be understood, actions defined in a respective event catch handler are executed. Alternatively, if a regular result has been returned, the result is used to fill one or more form items with values. In addition, form handlers which react on filled form items are executed.

Notable similar proposals and approaches include Speech Application Language Tags (SALT) [33] that targets multi-modal applications, and proprietary languages like the Philips Dialogue Specification Language (HDDL) [34], or the Generic Dialogue Modelling Language (GDML) [35] used in embedded systems in the automotive environment.

2.6 Information States and Dialogue Moves

Information state-based approaches present a compromise between the scripting approach described in the last section and the theoretically-inspired plan-based methods. A *dialogue move* is an abstraction similar to speech acts. It represents an action that a dialogue participant performs during the interaction. A *dialogue move engine* (DME, cf. [31]) is a system module that interactions with an information state in a rule-based way in order to integrate observed user moves, and to select a corresponding system move.

The system's basic cycle of operation (its *control algorithm*) consists of the following steps:

1. On the basis of the current information state, try to address obligations by performing and integrating the respective dialogue moves (*update* the information state), then yield turn.
2. Wait for user input by observing the user's dialogue moves and assume turn.
3. Integrate the user's dialogue moves and again update the information state.

Information-state based approaches have been used in a variety of projects. For instance, Matheson et al. [2000] have proposed several mechanisms related to grounding, i.e. the building and management of a common ground of beliefs between dialogue participants. The framework of *Questions Under Discussion* (QUD) has been proposed by Ginzburg [37]. Larsson has refined this framework, and developed an approach called *Issue-Based Dialogue Management* [31]. A sample structure of an information in this approach is shown in Figure 2.3. It may be regarded as a form of typed feature structure [38].

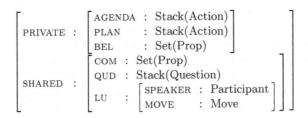

Fig. 2.3. An information state structure, cf. [31, p.36].

An information state represents the beliefs of a system (or participant) at some point in the dialogue interaction. Parts of the information state may be referred to using a path syntax, for instance, /SHARED/LU/MOVE. An information state typically consists of beliefs assumed to be shared (/SHARED) and those which are only held privately by the system (/PRIVATE), either because they have not yet been communicated or because they are only used internally. An information state framework provides the essential base types (for instance, stacks and sets) and operations required to access and operate on the structure.

During the interaction, the information is constantly updated. Typically, this takes place after a user message has been received and after the system has produced its response. The update takes place in terms of dialogue moves and update rules. The most well-known implementation of a dialogue move framework is probably the TrindiKit project [39]. In [40] an alternative formalisation and implementation is proposed.

An example of an update rule as used in TrindiKit is shown in Figure 2.4. When the update rule is applied to an information state, the preconditions specified in the PRE part of the rule are checked. The matching of these conditions may also have the (side-) effect of binding variables to values that can used in the EFF clause. For instance, this method is used in the update rule to bind the variable Q. Particular languages have been defined to express conditions and updates.

RULE : **integrateSysAsk**

CLASS : integrate

PRE : $\begin{cases} \$/\text{SHARED}/\text{LU}/\text{SPEAKER} == sys \\ in(\$/\text{SHARED}/\text{LU}/\text{MOVES}, ask(Q)) \end{cases}$

EFF : $\{ push(/\text{SHARED}/\text{QUD}, Q)$

Fig. 2.4. An information state update rule, cf. [31, p.43].

2.7 Concepts and Terminology

In the following, some common concepts and terminology from the area of dialogue systems development will be introduced, which will help to better understand the system descriptions in the next section.

Dialogue Initiative. The dialogue participant controlling the dialogue flow is said to have the *dialogue initiative*. Dialogue initiative has to be distinguished from turn taking. A dialogue participant who has assumed the turn may not have the dialogue initiative, for instance, when he assumed the turn just to answer a question raised by another participant.

In dialogue systems the following modes concerning dialogue initiative are distinguished:

- *System initiative* (directive mode): the system is in control of the dialogue and the system's utterances are designed to restrict the user's responses to small set of easily distinguishable options.
- *User initiative* (reactive mode): the user is in control of the dialogue flow, the system merely responds.
- *Mixed initiative*: both system and user may take initiative during the dialogue. The level of system initiative may still vary between *proactive* behaviour, only initiating clarification subdialogues, and a truly negotiated initiative.

Each of the modes has its own advantages and disadvantages depending on the context of use. For instance, novice users may be satisfied with a system-directed dialogue, because that strategy is more robust and guarantees in most cases that progress is made in the dialogue and that the chances for certain types of error such as misrecognitions may be minimised. A similar reason may apply to dialogue under adversary conditions such as noisy environments. Expert users, on the other hand, are more likely to prefer to assume the initiative in the dialogue, since they have a clear idea of what the system can be instructed to do. Technologically, allowing user initiative is more difficult that maintaining system initiative, because the system can establish fewer expectations concerning the next user utterance, which increases the chances of misinterpreting the user.

Dialogue Control. In dialogue systems, the term *dialogue control* refers to how the system's behaviour is implemented technically. Although this is an internal feature of the system, the kind of dialogue control chosen for the implementation determines to a high degree the kinds of interactions the system can engage in.

- Finite-state models offer a high degree of system control, which is vital in many environments. However, the limited expressiveness of finite-state models often incurs a certain inflexibility of dialogue systems implemented on their basis.
- Frame-based models extend finite-state models with more flexibility, especially concerning some aspects of mixed initiative (e.g. overanswering). A typical example is the VoiceXML form interpretation algorithm.
- Information state-based models are a general way of implementing dialogue strategies which may be used to implement finite-state, frame-based, or more complex models. Moreover, certain it has been used for implementing generic dialogue behaviours and strategies such as grounding and questions under discussion.
- Plan-based or agent-based models have been highly influential approach to modelling interaction and dialogue for a long time. An advantage of a plan-based model is that general behaviour rules may be stated which a dialogue system has to obey. Such rules of behaviour may include cooperation, helpfulness, sincerity. They may be, for instance, based on the Gricean maxims.
  However, very strong assumptions regarding the capabilities of the system and the setting have imposed severe limitations on the practical applicability of the framework.
- Statistical models and data-driven methods for dialogues are motivated by the success of these methods in other areas, most notably speech recognition. In principle, the idea of training a dialogue model is appealing. However, the limited availability of training data as well as the uncertainty as to the underlying representation whose parameters were to be trained has prevent widespread use of this approach to date.

2.8 Systems and Architectures

Since we aim at practical dialogue systems development, it is very instructive to review certain implemented SLDS and architectures. With the strong evolution of the field of dialogue systems, the list of systems to be discussed here is necessarily incomplete. For a more detailed presentation of a larger range of systems, we refer to reader to [3]. We focus on those that most clearly exhibit features that are relevant to our work.

Allen et al. [41, 42] introduce a dialogue systems classification scheme according to the following questions:

- Reference capabilities: Which referring expressions can be used in utterances? Possible options are static expressions, i.e. proper names or unique descriptions, and descriptions resolved by status of task. For instance, in

VoiceXML field-level grammars determine such a context. Alternatively, if personal pronouns in general is allowed, a general form of anaphora resolution based on salience and recency is required.

- Task complexity: Does the task consist of only a primitive action or is it composed of (hierarchically) structured actions? Or does the task merely consist of simple queries to a static database (such as in timetable information domain)? In addition, can multiple tasks be handled by the system, or can the task change as dialogue develops? In [43] the following additional questions regarding a plan-based domain are raised: What kinds of plan operations are possible? Is it possible to select and change different the recipe to use, including the temporal order of actions? Can the resources involved be changed? Is it possible to compare different alternative plans?
- Dialogue management capabilities: This concerns the question whether the dialogue itself can be used to manage the interaction. Thus, are certain phenomena and behaviours possible, such as clarification and error correction subdialogues? And if so, who can initiate them? In particular, if they can be only initiated by the system, its capabilities are restricted. Can the user or system or both answer with return questions? (e.g. "Do you mean X?"). Finally, is any form of "meta discussion" possible?
- Speech acts supported: This concerns the kinds of utterances that can be understood by the system. A basic system, for instance, may only produce questions and only accept assertions from the user. However, more complex speech acts such as those introducing obligations or promises are possible.
- Modality: This concerns the available modes to interact with the system. Typical modalities include speech, typed natural language, graphics, input from pointing or writing devices, as well as multimedia presentations on the output side. A key quality is also a system's ability to fuse input from different modalities and to produce aligned output presentations in different modalities.
- Incrementality: Is it possible to develop topics and utterances incrementally? Also, what forms of back channel responses and grounding are possible?

These questions may serve as a general framework for the discussion of the following systems and architectures.

2.8.1 Circuit Fix-It Shop

A pioneering spoken language dialogue system, called Circuit Fix-It Shop, is presented in [44, 45, 46]. It is relevant to our work because it presents one way to integrate dialogue management with a reasoning engine, in this case, a theorem prover based on Prolog [47].

As one of the earliest working systems, this system contained a speech recogniser with a non-trivial vocabulary. The domain of application is electronic circuit repair, with the system being an expert on the circuit workings, but without knowledge about concrete repair instances or sensing capabilities. Thus, collaboration and knowledge exchange is crucial in order to solve the common goal of diagnosing and repairing a certain circuit board. According to the authors the system implements essential behaviours required for efficient human-machine dialogue: problem solving to achieve a target, conducting and managing subdialogues, employing a user model, changing initiative during the interaction, and the generation of context-dependent expectations for speech recognition.

From the viewpoint of interaction experience, the system is interesting because it is able to operate in different system modes, termed *directive, suggestive, declarative,* and *passive*. In the directive mode the system is issuing command-like utterances, and user is expected to follow the instructions of the system. Also, interruptions and transitions to other subdialogues are restricted. The less directive modes provide more flexibility to the user. The system is also able to integrate a user model and evolve it during the interaction. The user model represents system beliefs concerning the knowledge of the user. For instance, whether or not the user knows where a certain knob is on the board. According to [45, p.4],

> The role of the user model is thus to supply or fail to supply axioms at the bottom of the proof tree. It both inhibits extraneous verbalism and enables interactions necessary to prove the theorem.

Technologically, the system is Prolog-based, but with certain extensions to the Prolog theorem proving approach in order to enable the interactive construction of a proof. To this end, the so-called *missing axiom theory* is designed to answer the question when and why the system should initiate interaction with the user. The behaviour of the system is driven by a theorem proving procedure that tries to prove a stipulated goal. When the system detects that the proof cannot be completed due to missing information it resorts to interaction with the user as an attempt to gather that information.

Internally, a recursive procedure called *ZmodSubdialog* is responsible for proving a given goal argument in the style of Prolog. Three main cases are distinguished in ZmodSubdialog concerning a goal R: If R is trivial, i.e. it can be executed by the system in a single step or it can be proved by the user model, the system assumes that this goal has been achieved and the proof proceeds. If R is a *vocalize* goal, the goal is achieved by outputting its content as a spoken utterance. Depending on the rule base, the respective goal may require the user to signal understanding and the performance of the indicated command or request. Otherwise, R is a complex goal with subactions defined in the knowledge base. In this case, new instantiations of the ZmodSubdialog procedure are created with the appropriate subgoals. More than one instance of a ZmodSubdialog call can be instantiated at one point in time. This represents different subdialogues that can be entered. If a subdialogue is exited, the actions in ZmodSubdialog are suspended, but may be resumed at a later stage in the interaction. The dialogue control module decides with which sub-dialogue to proceed.

In the case of a vocalize goal, the user can respond with different replies. In the case of an affirmative response ("Okay", "I've done it." etc.), the respective goal is proven and the proof procedure can continue. Otherwise, if the user indicates failure to perform or understand the action, the respective proof goal and dependent parts of the proof fail. This usually initiates backtracking in the proof search. Alternatively, the user can initiate clarification subdialogues: The system handles requests for clarification by "dynamically modifying the rule and re-executing with a newly required clarification subdialog." Finally, if the user responds in a way that is not expected by the current subdialogue, a transition to a different subdialogue is possible. Control is passed to other subdialogue if the user utterance is adequate to that subdialogue's current state.

Another interesting aspect of the system is the modelling employed in the knowledge base: it contains Prolog rules which are classified in three different categories: *General debugging rules* describe procedures in the circuit repair domain, for instance, how to "set a knob to a value". Typically, these rules include descriptions of higher-level goals in terms of subgoals. *General dialogue rules* represent the system knowledge about what can be achieved with vocalize goals, and, in particular, which goals can be vocalised. Finally, the *user model rules* describe the system's beliefs about what the user knows and what actions the user is capable of performing.

2.8.2 TRAINS and TRIPS

The University of Rochester's TRAINS [48, 49] project and its successor TRIPS [11] are efforts to build interactive planning systems. They feature spoken language interaction in application domains such as logistics, evacuation and emergency planning, and personal assistance, such as medication advice.

In order to develop the TRAINS system, human-human dialogues were recorded that feature rich problem-solving interactions in a logistics domain [50]. Typical problems included building transportation plans to move a certain amount of commodities to a specific location under time constraints. Although the domain is limited to a toy world as far as the number of locations, for instance, is concerned, the kinds of complex problem-solving phenomena that became apparent were highly complex. The actual TRAINS system contained a more limited domain in order to remain implementable. The modelling of the TRAINS domain and aspects of reasoning therein will be described in more detail in Chapters 4 and 5.

Concerning the interaction with the system, the user and system assume asymmetric roles. The user is responsible for the overall scheduling of actions, whereas the system should be supportive [51, 52]. In particular, the system may fill in details (that do not require the user's attention), suggest solutions, present and describe the current state of the world and how the current plan may affect it, and dispatch plans to external *agents*. These agents represent the system's facilities to interact with the physical world. The system needs to interpret reports from these agents, and, if necessary, coordinate the correction and modification of plans.

For its internal processing the TRAINS system is based on so-called *Episodic Logic* (EL) [53, 54]. This highly expressive logic is modelled to be closely related to natural language while at the same time being able to serve a practical reasoning tool. For instance, EL includes features like generalised quantifiers, lambda abstraction, sentence and predicate nominalisation, intensional predicates (for expressing wanting, needing, believing etc.), and underspecified representations (for instance, for reference resolution). The main uses of EL within the system are the representation of natural language utterances as obtained by a natural language parser, the modelling of the conversational interaction, and the representation of domain-level information in the form of event-based temporal logic. For these purposes and the processing of these representations, the RHET knowledge representation system has been applied.

TRAINS' successor TRIPS, for The Rochester Interactive Planning System, is a system that substantially generalises and extends the TRAINS approach. TRIPS has been applied to different domains, such as evacuation and medical assistance planning and medication advice, as well as in the military domain [55]. To this end, the TRIPS architecture incorporates a module called *Problem Solving Manager* which can be adapted to different domains. It is responsible for the domain-level problem solving operations such as building and evaluating plans. In the Pacifica evacuation domain, for instance, a human evacuation planning expert and the system collaborate to construct a complex transportation plan that has to fulfil various temporal and resource-dependent constraints. The architecture proposed by [56] is a refined version of the TRIPS architecture and is illustrated in Figure 2.5. It aims to address shortcomings of a so-called "pipelined" architectures in which the processing of user input essentially occurs in a serial fashion. Such processing implies that much time is spent waiting and, perhaps more severely, many translations between representations has to take place. According to the authors, this property limits the usefulness of conventional architectures for mixed initiative. In their proposed architecture, on the other hand, both the user and the system are explicitly treated as agents.

In the architecture proposed the main components, the Interpretation Manager, the Collaborative Agent, and the Generation Manager, operate asynchronously. For instance, the architecture provides the possibility for incremental interactions by enabling the interleaving of utterances with acknowledgement without assuming the user.

Concerning the interaction with the TRIPS system, certain interesting functionalities have been realised, most notably the detection of conflicts in planned activities and the exploration of scenarios using hypothetical reasoning. Domain-level conflicts may occur, for instance, when a bridge used

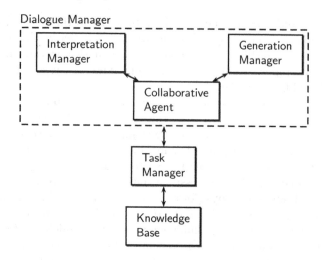

Fig. 2.5. Collaborative SLDS architecture proposed by [57, 56].

in an operation becomes unavailable. In that case, the planned actions affected, transportation events, are marked to indicate that they "may need to be revised". However, even though conflicts in certain actions are detected, it seems the user does not obtain a significant amount of support from the system concerning the decisions how to resolve the conflict.

Hypothetical reasoning is illustrated in the following slightly adapted dialogue fragment from the TRIPS system in the evacuation domain:

```
U > What if we went along the coast, instead?
S > That option would take 10 hours and 42 minutes.
U > Forget it.
```

In this dialogue, the user enters a hypothetical reasoning interaction (signalled by the key word "instead") in order to evaluate a specific scenario that differs from the current plan. This leads a new (and increased) system estimate for the overall time required to complete the planned activities. Consequently, the user cancels the exploration of this scenario.

Regarding the domain level on which these interactions are based, the TRIPS system is able to handle substantially more complex tasks than the TRAINS system. In the Pacifica domain, for instance, different types of transportation methods, each with specific constraints, are modelled. For instance, instead of ground transportation, helicopters may be used, but these are more limited in terms of capacity.

Interestingly, and in contrast to the earlier TRAINS system, the information about planned activities seems to be gathered mainly by performing stochastic simulation. This has advantages and disadvantages. On the one hand, it can be more realistic and to some extent more expressive than, for instance, a logic-based representation. The simulation of a plan may yield a probability distribution for certain parameter values like the duration. On the other hand, simulation limits the declarative notion of the modelling in the sense that the reasons for certain consequences, as in the dialogue example above, may not be traced back to the fundamental assumptions that implied it (i.e. that the longer duration is caused by the alternative route). Furthermore, the reliance on simulation may limit the ability of the system to deal with partial information because it may not be sufficient to perform a simulation.

2.8.3 COLLAGEN

COLLAGEN [58, 59, 60] is an application-independent *collaboration manager* based on the SharedPlan theory of discourse [61, 24]. The collaboration manager's task consists of assisting a human user in working with a domain application on a specific task, for instance, sending e-mail messages, programming a VCR, or planning a flight itinerary. COLLAGEN uses a discourse interpretation algorithm based on plan recognition [62, 63] in order to match user actions to instances of domain-specific recipes and integrate them into a shared interaction history. The dialogue is modelled in terms of Sidner's artificial discourse language (cf. Section 2.4). This theory is based on the ProposeForAccept and AcceptProposal primitives.

According to the authors, a key benefit of the domain-independent collaboration manager is the automatic construction of an interaction history that is hierarchically structured according to the user's and agent's goals and intentions. The goal of the collaboration manager in COLLAGEN is to achieve user success in working with some domain application that is accessible through its own direct-manipulation graphical user interface. The user is assisted by the COLLAGEN's *interface agent*, a software module that incorporates some knowledge about the application. Both the user and the interface agent are allowed to work directly with the application user interface. They perform actions such as keyboard or mouse input. They may, for instance, press one of the application's button controls or type in some text into a text field. These actions are observable to the respective other participant. The collaboration manager maintains and updates a plan structure that is derived from the recognised user goals based on a library of domain-specific plan templates, or *recipes*. The plan instance serves as a basis for the generation of suggestions and questions to be presented to the user.

In terms of the approach, COLLAGEN is substantially different from the systems discussed so far. These assumed a central dialogue management component to be in control of the interaction. In particular, it does not provide a general natural language input channel to the user. Instead, a menu-based facility to select possible next actions is proposed. In the COLLAGEN approach, the user is free to ignore the assistance offered by the system and work directly with the domain application. Also, COLLAGEN does not offer an alternative to replace the interface to the domain application functionality. Instead, it aims at assisting the user in interacting with the existing interface. As such, it bears some resemblance to a tutoring system.

According to Rich and Sidner, the development of an interface assistant for a different application requires only a relatively small additional amount of programming work. The work mainly relies in implementing a COLLAGEN-specific interface layer that translates between recipe steps and concrete application actions such as calling an executable function. An extension and re-implementation of the COLLAGEN approach is the DiamondHelp framework [64]. This Java-based framework emphasises the need for a modular and extensible architecture reusable across domains.

2.8.4 SmartKom

SmartKom [65] is interesting because of its approach to coherent multimodal interaction and distributed modular architecture. It distinguishes between several key components, such as discourse modelling, action planning, and the access to devices and services. SmartKom has also pioneered research into deploying an interface technology in substantially different usage environments, including its usage in a mobile context [66, 67].

On the input side, SmartKom realises a process called *media fusion*, such that user input from different modality streams can be tightly coupled. This is useful in particular for a mutual disambiguation of the input streams. For instance, a user utterance of the form "I want to go there" and a synchronous pointing gesture on a map display can be analysed together in order to resolve the referent of the phrase "there". In addition to these functionalities which the user can consciously take advantage of, the system attempts to recognise signals sent more subconsciously by the user, such as facial expression to indicate pleasure or anger.

In multimodal dialogue systems, utterance presentation is generalised to a coordinated generation of multimedia presentations. In the SmartKom project, an XML-based multimedia presentation language similar to SMIL has been developed and applied. The goal of this formalism is to achieve timed coordinated and efficient multimedia output, consisting, for instance, of avatar movements including gestures and lip movements, list and map presentations, as well as output of synchronous speech synthesised speech.

The component architecture of the SmartKom system is illustrated in Figure 2.6. Apart from the modality-related functionalities already mentioned, the architecture addresses important aspects related to our work. In particular, on the domain level SmartKom introduces a module called *function modelling* which is responsible for communicating with external task-specific functionalities, such as programming a VCR or telephone. The functionality commonly referred to dialogue management is spread via several tightly interacting components. The central role is assumed by the *action planning* component which determines the system's (inter-)action goal. To this end, it consults several components here subsumed under the label *knowledge services* which deal which include models of context, discourse, and the overall interaction.

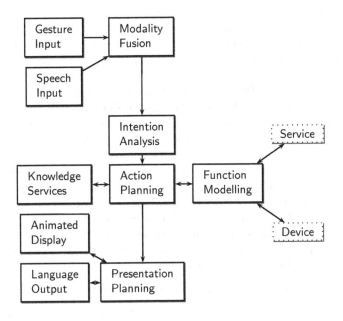

Fig. 2.6. The SmartKom component architecture [66].

The task of the function modelling is to determine which external devices are necessary and how they have to be controlled in order to handle a complex request like "Record the film XYZ tomorrow" in a plug-and-play fashion. The component employs a two-layered functional model consisting of an *internal* and an *external* model. The latter provides information about which functionalities are provided by a device that are relevant externally (to the user). The internal model, on the other hand, includes specialised information encoded in finite state machines that allow the function modelling component to control the device in order to perform specific operations. In this two-layered model, the recording request mentioned above is modelled as involving two devices, the actual VCR and a virtual scheduling device that triggers the action at the specified time.

The function modelling component is clearly related to our goal of integrating application domains. However, in contrast to our approach, the tasks realised within the SmartKom project seem to be tightly focused on the control and operation of external devices.

2.8.5 D'HOMME

Bos and Oka [2002] present an inference-based approach to dialogue system design and its realisation as a prototype in a "smart room" environment. D'Homme is particular in the sense that inference in First-Order Logic plays a central role in both the utterance interpretation and the system's decision making concerning responses and external actions. As such, the approach is similar to the one presented in [69]. The dialogue modelling is based on Discourse Representation Theory (DRT) [70, 71] which lends itself suitably to a translation into First-Order Logic.

The authors argue for the use of First-Order Logic to capture natural language meaning due to the expressiveness of FOL, its well-understood formalism, and the availability of promising tools. In particular, logical inference can contribute to ambiguity resolution in discourses, for instance, concerning the resolution of anaphoric expressions.

Their inference-based approach is based on the analysis of the complete dialogue, rather than a single utterance. The aim is to find a *consistent* semantic representation capturing the meaning of the dialogue. In particular, the failure to find such an interpretation is considered to signal the presence of problems such as misunderstandings between the dialogue participants.

In the following we will discuss some aspects of the approach in more detail. Consider the following user utterance:

U > Switch every light in the kitchen on!

This is translated by a natural language parser into the discourse representation structure (DRS) which is illustrated in Figure 2.7. In this DRS, the δ operator and the "t:" syntax are examples of the extensions to standard DRT which are used to represent actions and situations.

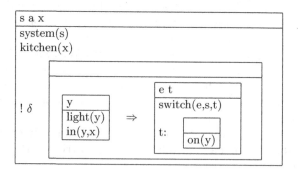

Fig. 2.7. Discourse Representation Structure of an instruction.

The DRS is subsequently translated into a formula of First-Order Logic. One may note that each predicate in the DRS is extended with a possible world referent. This approach emulates the use of modal logic by a first-order approximation.

$$\exists w, s, a, x \ . \ (possible\_world(w) \wedge system(w, s) \wedge kitchen(w, x) \wedge$$
$$\exists v, a \ . \ (action(w, a, v) \wedge \forall y \ . \ (light(a, y) \wedge in(a, y, x) \rightarrow$$
$$\exists e, t(switch(w, e, s, t) \wedge on(t, y)))))$$

The model-theoretic interpretation of this formula does not distinguish between possible worlds and (regular) entities in the domain. Thus, standard first-order inference tools can be applied. The authors also present an update algorithm that defines how complete dialogues can be interpreted using their approach. One of the central decisions that the algorithm has to take concerns the detection of the potential inconsistency of the user input. In the update algorithm, two reasoners are used in parallel. A model generation procedure is used to build models in case the input is satisfiable. At the same time, a theorem prover tries to prove the inconsistency of the problem. Whichever reasoner is first to return a result, will stop the other engine. If the theorem prover is successful, the input will be rejected. If, on the other hand, the input is satisfiable, it will be integrated into the evolving discourse context. In particular, if new entities denoting system actions are inferred, these actions will be executed.

The design presents an innovative approach to a logic-based architecture of dialogue systems. In particular, the flexibility of using a logic-based approach has been shown. However, the following issues present possible areas of improvement. Firstly, the model generation procedure employed is a black box. No proof structures are generated or used. It seems unclear how the system deals with multiple models. In addition to the model generator, a theorem prover for the case of inconsistent specifications is required. So, two different reasoners are used rather than one, which is undesirable. Furthermore, inconsistencies are seen as misunderstandings. However, there may be a lot of other reasons for logical inconsistencies, such as, the specification of an instruction that the system cannot fulfil. Finally, the system does not implement an incremental inference mechanism. This limits its applicability to small dialogues.

2.9 Discussion

This chapter has provided an introduction to some of the fundamental concepts and developments in the field of discourse and dialogue research. Concerning relevance to our approach, the main result of the theoretically motivated plan-based approaches is the definition of a domain level of representation. This level is used for representing the specific tasks that the user wants to accomplish. In such a task-oriented setting, they are the reason for the user to engage in a dialogue either with a human expert or a computer system.

In our approach to dialogue management, we do not follow the tradition built on the plan-based models. These have been criticised for being overly complex and rather abstract in their focus on the underlying intentions between utterances and "neglecting what the dialogue is about," namely the relation to the actual surface level. The diametrical direction of the dialogue scripting approach focuses on how to build actual systems that can be interacted with by actual users, in particular via speech. This focus stems not least from a commercial interest to provide services in the modality of spoken language. We view our approach based on domain-level reasoning as a building block in a compromise between these directions. We aim at balancing the necessary level of abstraction on the one hand with the desire to implement systems for specific applications on the other hand.

The systems discussed in the previous section show interesting aspects of interaction and dialogue behaviours concerning problem solving in task-oriented domains. However, none of the systems provides a set of functionalities that subsumes our goals of interactive reasoning and domain integration while maintaining a practical level of complexity.

For instance, the Circuit Fix-It system demonstrates how practical problem solving interactions can be modelled with the help of a theorem proving approach, i.e. in a logic-based manner. However, we do not follow this specific direction, since in that approach the theorem proving is applied on the level of the dialogue management, i.e. the different levels of representation concerning the dialogue and the domain are mixed. In our approach the reasoning engine operates exclusively on the domain representation. In particular, we aim at a clear separation of the domain knowledge from the interaction (and dialogue management) knowledge.

The COLLAGEN and DiamondHelp systems present an interesting approach to providing assistance in various application domains, such as travel planning or the control of home appliances. However, to our conception, the approach relies fairly much on plan-based metaphors and constructions. For instance, extensive plan recognition capabilities are required by the system to detect the intentions of the user. In addition, it seems unclear if transparent reasoning and an explanation of the system's behaviours can be achieved within that framework.

This criticism also applies to some extent to the TRAINS and TRIPS systems, although TRIPS is perhaps the system that pioneered hypothetical reasoning in the sense of "what if" questions in spoken dialogue. The system also provides mechanisms to detect conflicts in planned operations. However, in both cases the information seems to be provided by a simulation module, rather than by inference. In our opinion, this may be useful for obtaining the information, but restricts the ability to reason about the dependencies and thus limits the system's support for the user to make the right decisions, for instance how to resolve a conflict.

The D'HOMME project illustrates the flexibility that can be achieved by a consequent use of logic formalisms concerning natural language input and dialogue analysis. For instance, the use of natural language expressions including quantifiers has not been addressed in the other systems discussed here. However, from an architectural perspective the need to integrate two different reasoning engines and their use as autonomous black boxes seem to be candidates for improvement. In particular, in our approach we aim at a tight integration with the reasoning engine that will also be able to provide meaningful proof structures for its inferences. Furthermore, the domains addressed seem to be fairly limited. Thus it is difficult to say if the flexibility provided at the input level is reflected in the domain representation and its processing.

The SmartKom project has also been influential to our approach, also through the personal involvement in the project. From the perspective of domain functionality and usage scenarios, SmartKom has pioneered a movement toward system architectures that can be used in many environments and under different conditions. Also, a large number of services, such as the operation of home appliances, has been implemented on that level. Nevertheless, rather than on integrating these domains on the back-end in the sense that they can contribute to an overall task of the user, the SmartKom's strengths rely in its provision of a multi-modal front-end with novel and powerful features, such as a systematic modality fusion and fission.

3

First-Order Logic

3.1 Introduction

This chapter provides an introduction to First-Order Logic (FOL) in terms of general concepts and notations as well as in terms of its use as a modelling and inference device. As such, FOL is the basis of both our proposed domain modelling approach (cf. Chapter 4) and our implemented reasoning engine CIDRE (cf. Chapter 5).

We start by introducing the fundamental concepts of FOL in Section 3.2. A more detailed introduction and reference to First Order Logic may be found in [72]. First-order logic can be approached from two complementary perspectives, namely syntax and semantics: The syntax of FOL describes how well-formed formulas are constructed as symbol sequences. In addition, deduction (cf. Section 3.3) is a syntactical process that defines how these symbol sequences can be used to derive new formulas using rewrite rules. The semantics of FOL, on the other hand, is concerned with the formal definition of the meaning or interpretation of formulas in terms of logical models (cf. Section 3.4). In Section 3.5, we present Allen and Ferguson's approach to modelling actions and events in logic [73]. As part of our own approach, we adapt this theory to the specific requirements when used with a model generation procedure (cf. Chapter 4). We review approaches to model generation in Section 3.6, with a focus on tableaux-based methods. The first of these is the so-called Positive-Unit Hyper Resolution (PUHR) method. In fact, our proposed reasoning engine may be considered an interactive model generation procedure that is inspired by this method of inference. The Extended Positive Tableaux (EP), an extension of the PUHR method with more advantageous theoretical properties, is also discussed.

In our general approach proposed in this work, we will use First-Order Logic as a formalism to model and reason about information on the application domain level of a dialogue system. The aim here is to be able to derive inferences on instances of problems. This is to say we assume that typically the user will provide information in the form of ground statements. This information has to be processed using an axiomatisation of the application domain in order to derive consequences. We argue that First-Order Logic is an adequate tool for this task and we present the fundamental concepts that our approach is built upon in this chapter.

3.2 First-Order Languages

First-Order Logic can be seen as an extension of propositional logic (PL). PL only deals only with complete propositions, i.e. indivisible statements that are mapped to truth values. PL studies how these truth values can be combined by logical connectives. FOL, on the other hand, also studies the internal structure of statements in terms of individuals and the relationships that hold between them. Individuals are objects in the language (logical constants) and relations between them are represented by predicates. The essential difference between PL and FOL is that FOL can make use of quantification, i.e. it is possible to make statements about "all" or "some" individuals in a domain.

We will first define the language of first-order logic, i.e. the syntax of terms and formulas. Usually, when first-order logic is applied in a particular domain, the first step is to define the application-specific *vocabulary* (or *signature*). It is composed of a set of *predicate* symbols and a set of *function* symbols. Each symbol is associated with an *arity* $n \geq 0$. The arity of a predicate or function denotes the number of arguments the predicate or function takes. Conventionally, for each symbol there is exactly one arity defined. The notation f/n is used to state that f is of arity n. *Constants* can be considered functions of arity 0. These symbols are also called non-logical because they are not part of the logic itself, but rather "user-defined." The equality predicate "=" is not non-logical because it has special meaning and is interpreted specially in first-order logic.

In order to construct first-order terms and formulas, all first order languages make use of the following kinds of logical symbols:

- *Connectives*: The connectives ¬ (negation), ∨ (disjunction), ∧ (conjunction), → (implication) are used to construct complex formulas out of simpler ones. The usual syntactic convention for unary connectives is to write them in prefix notation, whereas binary connectives are written in infix notation.
- *Variables*: A countable set of variable symbols. The letters x, y, z etc. and indices are frequently used to refer to variables and to distinguish them from other terms, such as function symbols.
- *Quantifiers*: In FOL there are two quantifiers: ∀ for *universal*, and ∃ for *existential* quantification. Universal quantification is used to construct statements about all individuals in the domain, while existential quantification states the existence of at least one individual with a given property. Logically, existential quantification can be expressed by universal quantification and vice versa.

- *Parentheses*: Parenthesis symbols (,) are part of the language, they can be placed around symbol sequences as punctuation marks.

The non-logical and logical symbols together form the *alphabet* of the language. Often, the special predicates $\top$ and $\bot$ are also included as logical symbols, to denote truth and falsity, respectively, in all interpretations. Alternatively, they can be treated as syntactic sugar for $(P \vee \neg P)$ and $(\neg\top)$, respectively, for some arbitrary predicate P (as soon as well-formed formulas are defined).

We define the terms of FOL in the following. Terms are used within formulas to refer to individual objects about which a formula expresses a certain statement. The *terms* of FOL are constructed by the following inductive definition:

- Every constant from the set of constant symbols is a term.
- Every variable from the set of variable symbols is a term.
- Given a function symbol f, its arity $n > 0$, and a sequence of n terms $t_1, \ldots, t_n$, the application of f to the terms, $f(t_1, \ldots, t_n)$ is also a term.
- Nothing else is a term.

Based on the definition of FOL terms, the set of well-formed formulas of FOL can be constructed. First, atomic formulas are defined as follows: Given a predicate symbol P, its arity n, and a sequence of terms $t_1, \ldots, t_n$ as defined above, the application of P to the terms, $P(t_1, \ldots, t_n)$, is called an *atom* (or *atomic formula*). If ϕ is an atomic formula, then both ϕ and $\neg\phi$ are called *literals*. The equality predicate $=$ is written in infix notation between the terms.

Well-formed formulas (wff) can now be defined on the basis atomic formulas using the following formation rules:

- Any atomic formula is a wff.
- Negation: Given a well-formed formula ϕ, its negation $\neg\phi$, is a wff.
- Connectives: Given wff ϕ_1 and ϕ_2, $\phi_1 \wedge \phi_2$, $\phi_1 \vee \phi_2$, $\phi_1 \rightarrow \phi_2$ are wff.
- Quantification: Given a wff ϕ and a variable x, $\forall x . \phi$ and $\exists x . \phi$ are a wff.
- Parentheses can be placed around wff to form new wffs.
- Nothing else is a compound wff.

Given a well-formed formula, the notions of *free* and *bound* variables can be defined. Informally, variables that are under the scope of a quantifier are called *bound*, otherwise they are called *free*. A formula without free variables is also called a *sentence* or *closed*, otherwise it is called *open*. A formula without any variables is called *ground*.

A *clause* is a set of literals, implicitly interpreted as a disjunction. Due to the logical equivalence, we also use the representation (*implication form*) $\phi_1 \wedge \ldots \phi_n \rightarrow \psi_1 \vee \ldots \vee \psi_m$ for clauses of the form $\{\neg\phi_1, \ldots, \neg\phi_n, \psi_1, \ldots, \psi_m\}$. In implication form, the set $\{\phi_1, \ldots, \phi_n\}$ is called the *body* of the clause, and $\{\psi_1, \ldots, \psi_n\}$ is called its *head*. A *Horn clause* is a clause that contains at most one element in its head. We also identify a clause in that notation with its universal closure, i.e. with $\forall$ quantifiers added for each free variable in the clause, when the intended meaning is obvious from the context.

On the basis of well-formed formulas, a few auxiliary notions can be introduced. A *substitution* $\sigma = [t/x]$ is a function that, applied to a formula ϕ, replaces all free occurrences of the variable x by the term t. This is written as $\phi\sigma$ or $\phi[t/x]$. This notation be generalised to the case when finitely many variables $x_1, \ldots, x_n$ are replaces by associated terms.

Another important concept is *unification*: A substitution σ is a *unifier* of a set of formulas $\Delta = \{\phi_1, \ldots, \phi_n\}$ if and only if for all $1 \leq i, j \leq n$ holds $\phi_i\sigma = \phi_j\sigma$. In this case, the set of formulas is called *unifiable*. σ is called a *most general unifier* (mgu) if and only if , intuitively, every other substitution may been represented as an extension of σ. The idea behind most general unifiers is that only those substitutions are applied that are necessary to make the formulas in Δ equal. In a similar way, it is also possible to generalise the notions concerning unification to terms instead of formulas.

3.3 Deduction

Having defined the syntax of well-formed formulas in the language of first-order logic, we can define a way how these formulas can be processed. *Deduction* is the syntactical process of deriving new formulas from given ones. It is based on the syntactical properties of these formulas and, in contrast to the model-theoretic semantics discussed later, not on how these formulas are interpreted. Based on the definition of well-formed formulas, the *calculus* of first-order logic consists of *inference rules* and *axioms* (also known as its *proof theory*). Inference rules describe how formulas can be constructed from given ones. An inference rule consists of a set of *antecedents* (formulas that have to hold), and one *consequent* formula (a formula that is the result of the inference rule).

Inference rules are written in the following notation:

$$\frac{A1, A2, \ldots, An}{C}$$

Axioms are formulas that are taken to hold *per se* (as assumptions, without proof). One may distinguish *logical* axioms from *non-logical* axioms: logical axioms are "built-in" into FOL in the sense that in every FOL language these axioms are required to hold. Non-logical axioms define some "application-dependent" first-order theory. The Peano axioms, for instance, define the properties of natural numbers. First-order logic can be equivalently characterised by many different combinations of inference rules and axioms. Typically, the more inference rules are allowed, the fewer axioms are required, and vice versa. *Axiom schemata* are used to specify calculi in which an infinite number of axioms are used.

Based on the definition of the language of FOL, the notion of a *deductive system* D can be introduced. A deductive system defines derivations of new formulas, based on the particular inference rules used, which can be said to be *deductively valid*.

An *argument* is a non-empty collection of formulas. These formulas are partitioned into one *conclusion* and a set of *premises*. This is written as $< \Delta, \phi >$. $\Delta \vdash \phi$ means that ϕ is *deducible* from Δ. This is equivalent to stating $< \Delta, \phi >$ is deducible in D. A set of formulas Δ is *consistent* if and only if there is no formula θ which is proved by Δ, as well as the negation of θ.

Resolution is one important inference rule. The idea of resolution is that given two clauses (disjunctions of literals) where one literal from the first clause appears in negated form in the second clause, both literals can be removed from the disjunctions to form a new disjunction, the so-called *resolvent*. This is formalised in the following inference rule:

$$\frac{a \vee b, -b \vee c}{a \vee c}$$

In the propositional case resolution is a sound and complete method to decide the validity of a formula in PL. For the use in FOL it can be generalised using implicit quantification for $\forall$, Skolem functions for $\exists$, and Unification. Determining the validity of a formula in first-order logic is only semi-decidable in the general case.

3.4 Model-Theoretic Semantics

Deduction introduces one kind of correctness of a given formula in first-order logic (a proof for the formula can be constructed in predicate logic using deductive inference rules). Model-theoretic semantics is a complementary view to define correctness in terms of its *meaning* in a domain.

A *model* $\mathcal{M} = (\mathcal{D}, \mathcal{I})$ is a pair of a non-empty set $\mathcal{D}$, called the *domain*, and an *interpretation function* $\mathcal{I}$.

The interpretation function $\mathcal{I}$ maps each n-place function symbol into a n-ary function over the domain $\mathcal{D}$; this includes constant symbols. It also maps each n-place predicate symbol to an n-ary relation over $\mathcal{D}$ (a subset of $\mathcal{D}^n$)[1]. A *variable assignment* is a function over the domain $\mathcal{D}$ that maps variables to individuals in $\mathcal{D}$.

With these preliminaries, the interpretation of a term, given an interpretation function $\mathcal{I}$ and a variable assignment g, can be inductively defined. If the term is a variable, its interpretation is defined by the variable assignment. Otherwise, it is an n-ary function symbol f applied to some terms $t_1, \ldots, t_n$, in which case the interpretation is defined recursively by mapping f to a function over the domain and applying to the interpretations of the terms.

---

[1] Or, in an alternative formulation, to an n-place function $p : \mathcal{D}^n \to \{true, false\}$.

$$val_{\mathcal{I},g}(x) = g(x).$$
$$val_{\mathcal{I},g}(f(t_1,\ldots,t_n)) = f^{\mathcal{I}}(val_{\mathcal{I},g}(t_1),\ldots,val_{\mathcal{I},g}(t_n)).$$

Given the interpretation of terms, we can define in a similar way the truth value of a formula in a model $\mathcal{M}$.

Now the above definition of *satisfaction* can be formalised: A well-formed formula ϕ is *satisfied* by a model $\mathcal{M} = (\mathcal{D}, \mathcal{I})$ and an assignment function s, written as $\mathcal{M}, s \models \phi$, if and only if the following inductive definitions holds:

- $\mathcal{M}, s \models P(t_1,\ldots,t_n)$ if and only if $(val_{\mathcal{I},s}(t_1),\ldots,val_{\mathcal{I},s}(t_n)) \in P^{\mathcal{I}}$
- $\mathcal{M}, s \models \neg\phi$ if and only if it is not the case that $\mathcal{M}, s \models \phi$
- $\mathcal{M}, s \models \phi \wedge \psi$ if and only if $\mathcal{M}, s \models \phi$ and $\mathcal{M}, s \models \psi$
- $\mathcal{M}, s \models \phi \vee \psi$ if and only if $\mathcal{M}, s \models \phi$ or $\mathcal{M}, s \models \psi$
- $\mathcal{M}, s \models \forall x \,.\, \phi$ if and only if $\mathcal{M}, s' \models \phi$ for all assignments s' that differ from s at most in the assignment to variable x
- $\mathcal{M}, s \models \exists x \,.\, \phi$ if and only if $\mathcal{M}, s' \models \phi$ for some s' that differ from s at most in the variable x

If ϕ is a closed formula, it is independent of the variable assignment and thus one may write equivalently $\mathcal{M} \models \phi$. A closed formula ϕ is *satisfiable* if and only if there is a model $\mathcal{M}$ that satisfies it, i.e. there is an $\mathcal{M}$, such that $\mathcal{M} \models \phi$. A closed formula ϕ is *valid*, written $\models \phi$, if and only if it is satisfied in all models, i.e. for all models $\mathcal{M}$, $\mathcal{M} \models \phi$ holds. In that case it is also called a *logical truth*. Finally, An argument $< \Delta, \phi >$ is valid, or the formula ϕ is a *logical consequence* of the set of formulas Δ if and only if if there is no model in which all the premises in Δ are true, but the consequence ϕ is not true.

The Herbrand interpretation is a special form of interpretation function. It is used, for instance, in the PUHR tableaux method that is discussed in Section 3.6. The *Herbrand interpretation* is an interpretation function that maps each term to itself. Its range is the *Herbrand universe*, which, given a set of clauses S, is the set of terms that can be constructed out of all constants appearing in S using the function symbols appearing in S. In the case that S does not contain any constants, an arbitrary constant is to be used. This set is infinite, as soon as there is a non-constant function symbol. It can be shown that a set of clauses S is unsatisfiable if and only if there is no Herbrand model for S.

The definitions of soundness and completeness link the syntactic and semantic views of logic. The deductive system D is *sound*, if and only if any formula that can be deduced ($\Delta \vdash \phi$) is semantically valid ($\Delta \models \phi$). A deductive system D is *complete*, if and only if any valid formula can be deduced.

3.5 Modelling Actions and Events in Logic

This section reviews Allen and Ferguson's approach to modelling actions and events in interval temporal logic[2] [74, 73]. This approach is the basis of our proposed formalisation described in Chapter 4.

Allen and Ferguson argue for an explicit representation of time, especially in planning. They claim that predominant alternative approaches to planning cannot handle certain related problems without "dramatic extensions" that would amount to extending these approaches with an explicit model equivalent to theirs. The situation calculus (Hayes 1969), for instance, has problems representing the duration of an event, which is a deficiency that is shared by STRIPS [75] representation-based approaches.

The basis of Allen and Ferguson's approach is an interval-based representation of time in a "simple linear model of time" in which all intervals are finite, but there is no beginning or ending of time. Allen and Ferguson introduce six binary relations that can hold between time intervals. These relations are illustrated in Figure 3.1.

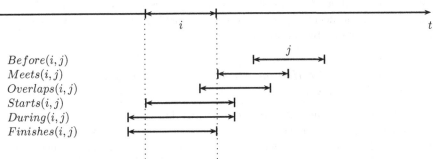

Fig. 3.1. Interval temporal relations according to [73]. This is to be read as $Before(i, j)$, i is before j, given the two intervals i and j as illustrated in the figure.

The axiomatisation of time makes use of the primitive predicate $Meets/2$, and consists of the following axioms:

- For every period there exists a period that it meets and another one that meets it.
- Periods can be concatenated to form new periods.
- The intervals meeting a certain period form an equivalence class.
- Intervals are equal if they are between the same periods.
- Periods, or more precisely, the meeting points of pairs of periods can be linearly ordered.

---

[2] The term *temporal logic* is used by the authors, however standard First-Order Logic is used.

Given the *Meets* primitive, derived predicates can be defined. For instance, two intervals i and j are in a *Before* relation if and only if there exists another interval m between them:

$$Before(i, j) \equiv \exists m \; . \; Meets(i, m) \land Meets(m, j)$$

This also covers the case when m is empty, i.e. i and j meet.

The use of intervals as the basic units of the logic implies that care must be taken in the definition of predicates. In particular, axiom schemata defining the *homogeneity* of a predicate and certain restrictions (*discrete variation*) concerning truth value assignments to subintervals are necessary.

The authors note that their logic is a standard first order logic and no extensions to first order logic are necessary to reason about the temporal aspects. However, some axiom schemata are proposed for different situations.

The definition and axiomatisation of time provides the ground for the actual core of the theory, the representation of events. Events are reified, i.e. they are regular individuals in the domain, and can be referred to using terms (Davidson 1967). This has been proved useful due to the potentially unbounded number of qualifications that may describe aspects of an event. For instance, the necessary conditions of a stacking action in a blocks world domain is described by the following axiom:

$\forall x, y, t, e \; . \; Stack(x, y, t, e) \rightarrow$
$Clear(x, pre1(e)) \land Holding(x, con1(e)) \land Clear(x, eff1(e))$
$\land \; Clear(y, pre2(e)) \land On(x, y, eff2(e)).$

In this formula, e is the reified event entity, while *Clear*, *Holding*, and *On* are properties that hold between blocks x and y at the respective time intervals. $pre1(e)$ and $pre2(e)$ are intervals that meet the interval $con1(e)$ during the event, which in turns meets the intervals $eff1(e)$ and $eff2(e)$ that mark the time after the event. This is intended to represent that when the Stack event is started, both blocks x and y must be clear, and block x will still be clear after the event, when it is on y.

When representing the effects of actions and events in this way, the *frame problem* arises [76]. That is, the problem that one not only has to specify (i) what kind of changes an event brings about in the world, but also (ii) which parts of the world state remain unaffected by the event. For instance, "calling someone on the telephone does not change the height of the Eiffel tower." Stating this explicitly is necessary because otherwise it cannot be decided if a certain property that is known to hold before an event still holds after the event. Typically so-called *frame axioms* are introduced to explicitly model this knowledge. However, this approach can be criticised, because the number of such axioms can grow beyond any practical size when a larger domain is modelled. A particularly undesirable consequence is that completely unrelated aspects of a model (as seen in the example above) need to be merged in these kinds of axioms. Alternative approaches to tackle that problem include using a default logic or abduction are mentioned, but not discussed in detail, because a different way to address the problem is presented.

The frame problem is addressed by Allen and Ferguson in a way similar to frame axioms in the sense that axioms are used to describe when events must have occurred. Therefore, these axioms are based on properties that change during some period of time, rather than describing what properties remain unchanged after each possible action. This approach limits the number of necessary frame axioms, and in particular, it is more modular than the traditional approach. For instance, in the Eiffel tower example above, it may stated that the height of the Eiffel tower is only changed by an earthquake. This would automatically imply that a simple telephone call alone cannot change the height of the monument.

Allen and Ferguson extend their theory of event representation to include a theory of causality. The cornerstone of that theory is the distinction between trying to perform an action and the actual occurrence of the action or event. This is reflected in their approach of using different axiom schemata for the different conditions: One part is used to describe what is necessarily true when an event has happened. The other describes what is sufficient for an attempt to be successful. According to the authors, the distinction makes sense because these two kinds of conditions are typically from different kinds of knowledge sources. For instance, one may know the conditions for a car repair event to have happened (the car was broken, now it is working again), while not knowing what exact conditions are necessary for a car repair attempt to succeed.

3.6 Finite Model Generation Using Tableaux

Model generation is the process of finding a model $\mathcal{M} = (\mathcal{D}, \mathcal{I})$ of a specification S, such that $\mathcal{M} \models S$. Finite model generation, in particular, is concerned with finding finite interpretations, i.e. ones that only contain a finite number of individuals in $\mathcal{D}$. In our case, as in the most applications, only finite models are meaningful.

Determining whether or not a model exists for a given specification is undecidable (semi-decidable) in general. However, if a theory is known to have a finite model, it can be found given enough time by enumerating and checking the candidates.

Model generation can be approached in two different ways. The first one is to try to enumerate all model candidates and checking them. The second one is to use a tableau-based method. The first method essentially consists of finding a model by fixing its maximal size and translating the specification into a propositional formula by replacing all quantifiers with their finite equivalent. This means that these methods reduce a FOL problem to a SAT problem in propositional logic, which is decidable. The size of the domain is increased if no model is found (iterative deepening). Systems implementing this technique are, for instance, FINDER and SEM [77]. Essentially, these methods are based on sophisticated techniques for enumerating all interpretations of a given size and checking efficiently if this interpretation satisfies all required formulas. There are a number of drawbacks of these methods, however. First of all, they cannot detect unsatisfiability, i.e. the inconsistency of the given theory. They can only report that no model has been found up to a fixed size limit. In addition, even though the respective algorithms are highly tuned for efficiency, the sizes of models that can be practically checked is relatively low. Finally, their reasoning is arguably less close to "common sense reasoning" and therefore less adequate for a human-machine interaction setting. For instance, these methods cannot directly detect which premises an inference relies on etc. However, this kind of information is essential in our approach.

Therefore, we concentrate on tableaux-based methods, and specifically a method called Positive Unit Hyper-Resolution (PUHR) [78, 79]. In this method, logic formulas are processed as clauses in implication form. The clauses have to be *range-restricted*, i.e. all variables in the head of the clause also appear in body.

Range-restriction is a relatively weak requirement in the sense that a given set of clauses which are not necessarily range-restricted can be transformed into a set of range-restricted clauses, without affecting satisfiability of the set and with only minimal modifications to the models. More precisely, any model of the original theory can be translated into a model of the range-restricted theory, by adding a new auxiliary predicate D that enumerates all the entities in the domain.

The PUHR method for model generation is based on so-called PUHR tableaux. A PUHR tableau for a given set of range-restricted clauses S is a tree structure that represents the search space in the sense that the branches represent potential models of S. The nodes of the tree are labelled with sets of ground atoms and disjunctions of ground atoms. In contrast to other methods, PUHR tableaux are built using only two expansion rules.

More precisely, a *PUHR tableau* T for a set of range-restricted clauses S is a tree where each node is labelled with a set of ground atoms or disjunctions of ground atoms which is either the unit tree $\{\top\}$ or is recursively constructed using the following PUHR tableau construction rules applied to a leaf in the tableau:

- *PUHR rule*: Given atoms A that unify with the preconditions of a clause C in S, resulting in a most general unifier μ, a new tree node is added as a child of N and is labelled with the clause's head, with μ applied to it. Since A consists of only ground atoms and C is range-restricted, the resulting formula is a ground atom or a disjunction of ground atoms.
- *Splitting rule*: Given a disjunction in the tableau, new child nodes, one for each element of the disjunction are attached to N.

If an (infinite) sequence of tableaux is constructed for a clause set S, such that the first tableau is the initial tableau, and in each step the tableau is transformed by one of the two rules above, then the union of this sequence of trees is an (infinite) tableau of S. It is worth noting that only positive formulas and no negative formulas are explicitly represented in a PUHR tableau.

A branch in a PUHR tableau is called *closed* if and only if its label formulas contain $\bot$, otherwise it is called *open*. In analogy, a PUHR tableau is called *closed* if and only if all branches are closed, or *open* if there is at least one open branch.

Figure 3.2 illustrates an example of a PUHR tableau. In particular, an infinite tableau can be constructed from the following set of clauses S:

$$p(a).$$
$$p(x) \rightarrow q(x) \vee p(f(x)).$$

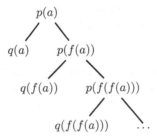

Fig. 3.2. The derivation of an infinite tableau. For the sake of readability, the tableau has been simplified by (i) combining the two PUHR construction rules in each level of the tree and (ii) showing only the newly added formulas in each node.

Informally, a tableau is *fair* if and only if all possible tableau construction rules have been applied. A branch A of a tableau is called *fair* if it is closed, or it is open and all possible applications of the tableau construction rules have been performed. Thus, a tableau is called *fair* if and only if all its branches are fair. An open fair branch is also called *saturated*. Fair branches can be finite or infinite.

The following important theoretical properties can be shown to hold:

- **refutation soundness**: Soundness with respect to unsatisfiability. If there is a closed tableau of S, then S is unsatisfiable.
- **refutation completeness**: Given an unsatisfiable set of clauses S, all fair PUHR tableaux are closed.
- **model soundness**: Given a satisfiable set S and a fair tableau, every open branch represents a model of S.

In order to be computationally more efficient, i.e. to avoid returning redundant or non-minimal models, the splitting rule can be adapted (called *component splitting*) and the tableaux have to be constructed using an ordered depth-first search strategy.

However, the PUHR method still has the following drawbacks:

1. It is not complete for finite satisfiability, i.e. it may fail to detect some existing finite models.
2. It requires Skolemisation, which is an obstacle for certain kinds of formulations.

In the following we discuss an extension of the PUHR method that addresses these drawbacks [80]. The essential difference is that this approach is complete for finite satisfiability, in addition to the properties of the PUHR method. This means that it is able to build models for theories like $\{p(a), \forall x . p(x) \rightarrow \exists y . p(y)\}$, whereas the original PUHR method would initiate an infinite model instantiation for the Skolemised version. Completeness for finite satisfiability is made possible by a different treatment of the existential quantifier. In particular, no Skolemisation or reduction to implication form is required.

Rather than using clauses in implication form (as the PUHR method), so-called "positive formulas with restricted quantifications" (*PRQ formulas*) are employed here: These have the same expressive power than standard first-order logic formulas [80]. Consider the following example of a PRQ formula:

$$\forall x . (employee(x) \rightarrow \exists y . (boss(y) \land works\_for(x, y)))$$

In this example, $employee(x)$ and $boss(y)$ are the ranges of x and y, respectively. Informally speaking, these ranges are used in the method to iterate over the quantified variable. In the case of $\forall$, this corresponds to range restriction.

The resulting tableaux method is called *Extended Positive Tableaux* (EP Tableaux). The key to EP tableaux is the definition of the extended $\exists$ (or δ^*) rule, where c_{new} is a term not used in the current branch:

$$\frac{\exists x . E(x)}{E[c_1/x] \mid \ldots \mid E[c_k/x] \mid E[c_{new}/x]}$$

What can be seen from the rule is that in the case of an existentially quantified formula all candidates for x are tried, as defined by its range, before introducing a new term.

This results in the **finite satisfiability completeness** for EP tableaux: they construct all the finite term models of a set of satisfiable PRQ formulas, up to a constant renaming.

3.7 Discussion

In this chapter we have provided an overview of important and well-known concepts of First-Order Logic. FOL will be the basis for our interactive reasoning approach. In particular, this concerns the following two aspects that have been introduced: First, a generic modelling theory to describe actions and events with special regard to their temporal structure and dependencies, and second, the concept of model generation using tableaux-based methods as an inference procedure.

Being able to model actions and events is a central requirement for many application domains. This will be further discussed in the next chapter. The usage of model generation as an inference procedure and the required adaptations for a use in an interactive environment will be the topic of Chapter 5. With regard to the different variations of model generation, we can note that FINDER-style approaches are too restricted in terms of the number of entities, and their reasoning can also be criticised for being non-intuitive. In particular, no proofs of conflicts are produced. Nor is it possible to directly make use of multiple solutions, if they are present. Thus, for instance, when a resulting model is inspected, one cannot distinguish between facts that are required (in all models) or suggested (only in some models). Instead, a tableaux-based PUHR method, can address these issues. The PUHR method is relatively intuitive and relies on a small number of tableau construction rules, which is desirable for both implementation and explanation uses. The EP tableaux method which refines and extends the PUHR method has better theoretical properties. However, in practice it may prematurely commit to certain decisions, i.e. variable instantiations. This may make the reasoning less adequate for interactive purposes, because to some extend "too much guessing" is required in the process. Another drawback is that its input format does not allow function symbols which are typically a useful ingredient in the modelling of domains.

We only briefly mention that First-Order Logic is not the only kind of logic that may be considered for representation and inference purposes. For instance, second order and higher-order logics introduce the possibility to quantify over relations, not just individuals, which increases their expressive capabilities. One may, for instance, define a second-order predicate *trans* that denotes the property of transitivity of a first-order relation, in the following way:

$$\forall P \; . \; trans(P) \equiv (\forall x, y, z. P(x,y) \wedge P(y,z) \rightarrow P(x,z))$$

These logics are substantially more complex than FOL, and in particular, there is no automated system that is sound and complete for second order logics and above.

Modal logics can be considered another extension of FOL. They exist is various different forms and introduce new logical operators. For instance, in temporal modal logic, the one-place operator $\Box$, corresponding to $\forall$, is introduced. It states that a formula is to hold always. The dual operator $\Diamond$, corresponding to $\exists$, denotes "at some time". All formulas are implicitly associated with a temporal annotation, the *world* in which they are to hold. Different worlds are connected by a transitive *accessibility* relation. If a formula is to hold always, it is to hold in the *current world* and all worlds that are accessible from the current one. Another variant is the modal logic of necessity and possibility. This is similar to the case of temporal logic. However, here the operator corresponding to $\forall$ states that "it is necessary that f holds". These logics can also be combined with temporal logics. In summary it can be stated that extended logical formalisms are much more complex in terms of processing and that their added representational capability may not be needed in the kinds of applications we have in mind.

4

Logic-Based Domain Modelling

4.1 Introduction

In this chapter, we describe our approach to a logic-based modelling of application domains. The modelling concerns structure of tasks that a user may want to achieve using a dialogue systems. We refer to these models[1] as *domain theories*. The task of the domain modelling can be broadly divided in two areas: The modelling of static information, i.e. structural information that does not change over time, and, on the other hand, the dynamic information which concerns time-dependent processes and changes in the states of objects. Of course, the two kinds of models are tightly interconnected in a typical application domain. Individual objects on the domain level, such as a geographical location or an appointment, are referred to as *entities*. Concerning the static information, the modelling of *classes* of entities can be distinguished from the modelling of their instances in a domain. The first problem is often referred to as the construction of a classification scheme, or ontology, while the latter has resemblance to representing items in a data base. Concerning the dynamic models, one of the most important aspects is the formalisation of so-called *fluents* and fluent change events. This approach is inspired by Allen and Ferguson's Interval Temporal Logic (cf. Section 3.5). However, it is substantially adapted and extended in order to be useful in combination with our reasoning engine CIDRE (which is described in Chapter 5). We mainly defend the usefulness of our modelling approach by describing how important concepts from different domains are generalised into modules to form a library of reusable domain theories (cf. Section 4.4). In particular, we discuss the modelling of the TRAINS domain on the basis of the concepts introduced.

At the syntactic level, the domain models we develop will essentially be clausal theories, i.e. set of First-Order Logic clauses, written in implication form. However, for brevity we sometimes use non-clause formulas which can be

---

[1] The more loose usage of the term "model" is not to be confused with the notion of a logical model introduced in Chapter 3.

straightforwardly translated into clauses, for instance $A \rightarrow B \wedge C$. In addition, for each n-ary function f we assume an $n + 1$-ary relation f adhering to the following definition:

$$f(t_1, \ldots, t_n) = x \equiv f(t_1, \ldots, t_n, x)$$

Note that confusion between the use as a function and the use as a predicate are usually avoided by the context.

4.2 Modelling Structural Information

In this section, we introduce a basic approach to capture classification schemes in a logic-based representation. These kinds of classifications are useful for structuring our library of domain theories which will be described in Section 4.4.

An *ontology* is a formal and explicit representation of a classification scheme for entities [81]. It provides a vocabulary that may be used to speak about classes, relations, functions, and other objects. In many cases, an ontology contains a hierarchical ordering of objects in super and subclasses. subclasses may inherit properties from their superclasses. In general, a class may have multiple superclasses. The subclass relation is a partial ordering (i.e. a class cannot be its own sub or superclass.) For instance, a real world example is the scientific systematics (taxonomy) which classifies biological life forms. Usually, such classifications are used for properties of objects that are static in the domain that is modelled. For instance, the quality of being a person, having a certain property, etc., does not change depending on time.

Simple class hierarchies can be straightforwardly encoded in a clausal theory. A class A with (a complete partition into) subclasses A_1, A_2, A_3 can be described in a logic theory in the following way: An element x of class A is either in A_1, A_2, or A_3 (or more than one, if the subclasses are not disjoint):

$$\forall x \; . \; A(x) \rightarrow A_1(x) \vee A_2(x) \vee A_3(x)$$

Any element x of the classes A_1, A_2, and A_3 is a member of class A:

$$\forall x \; . \; A_1(x) \rightarrow A(x).$$
$$\forall x \; . \; A_2(x) \rightarrow A(x).$$
$$\forall x \; . \; A_3(x) \rightarrow A(x).$$

If any subclasses A_i and A_j are disjoint, then no element x can belong to two different classes:

$$\forall i, j, x \; . \; A_i(x) \wedge A_j(x) \rightarrow i = j$$

If a relation is known to hold for the elements of a superclass A, then it is automatically known to hold for the elements of its subclasses A_i (since all elements of A_i are also elements of A). In order to realise multiple inheritance it is possible to generalise these equations for multiple superclasses.

One of the things that cannot be modelled as elegantly is an exception from inheritance, or *default inheritance*. For instance, birds usually (by default) can fly. However, some birds (such as penguins or ostriches) cannot. In a default inheritance scheme, one would associate the ability to fly with the base class and allow exceptions for certain subclasses. However, in our modelling approach these exceptions have to be made explicit when defining the base class.

We have implemented a simple scheme for translating a basic conceptualisation encoded as an XML document into a logical theory. Figure 4.1 shows an example application in the TRAINS domain. In the XML-based representation the sibling classes are assumed to be exclusive, except when an attribute complete="no" is included. The attribute entities can be used to specify a list of entities belonging to the class.

```
<class name="top" complete="no">
  <class name="transport−car">
    <class name="boxcar">
      <class name="boxcar1" entities="b1 b2 b3"/>
      <class name="boxcar2" entities="b4 b5"/>
      <class name="boxcar3" entities="b6 b7"/>
    </class>
    <class name="tanker"/>
  </class>
  <class name="track"/>
  . . .
</class>
```

Fig. 4.1. Simplified XML-based classification for the TRAINS domain.

Ontologies commonly also describe the structure of objects by the properties they have. These properties are usually modelled as functions mapping the object to a specific value of a given class. One approach to a declarative formalisation of a conceptualisation is the application of description logics [82]. Description logics are a restricted form of First-Order Logic. These provide ways to a more expressive representation of classifications than the basic approach presented here.

4.3 Modelling Dynamic Properties

Our formalisation is an adaptation and to some extent a simplification of Allen and Ferguson's theory presented in the Section 3.5. It is a simplification because it uses a point-based representation of time. For instance, we do not require Allen and Ferguson's homogeneity axioms, and can furthermore restrict the kinds of temporal relations that can arise. It is an adaptation because our approach introduces certain concepts that aim at making the theory suitable for a model generation procedure. It is also similar to the event calculus [83]. We also do not include a theory of causality in the form of Allen and Ferguson, because this does not seem to be essential in the domains we are focusing on.

We follow Allen and Ferguson in the following fundamental insights, that actions take time, that they may overlap or occur simultaneously, and that, if so, they may interact in complex ways. We are also inspired by their approach of using frame axioms based on properties rather than describing "non-effects" of events. Despite of our use of a point-based temporal structure, the fundamental approach is similar. In particular, we also adopt a Davidsonian-style, i.e. reified, representation of events.

The idea is to formally define logical consistency rules for events that, as an effect, change the state of the world in some way. More specifically, we focus on changes that are applied to dynamic properties of objects. For instance, travelling can be viewed as a location change, transporting objects implies a load change of the transporter (loading and unloading the objects). In addition, working with abstract objects such as a calendar or a reservation system, the meeting time may change, and other properties, such as location. Performing reservations (e.g. in a hotel, or in a cinema) can be seen as a change in the allocation of some object (hotel room, cinema seat) to some person(s).

Allen and Ferguson discuss the changes in properties that are brought about by the occurrence of some action or event. A property in their terminology is a logical statement that holds (or does not). In our approach, we use the term *property* to refer to a specific value-based quality of an entity in the domain. Thus, in our approach, properties are binary functions which map a pair (x, t) of an entity x and a time-point t to a value that also is an entity in the domain. Consequently, the frame axiom-like rules we consider in our approach take the form of the following scheme, where x is a domain entity, s and t are time-points and EV_i stands for the occurrence of some event of type EV_i.

$$change(x, s, t) \rightarrow EV_1 \vee \ldots \vee EV_n$$

Here and in the sequel, for the sake of readability we omit universal quantifiers, since we do not refer to constants.

Modelling time. Time can be modelled in very different ways, depending on the aims of the modelling [84]. For instance, the model can be either linear or branching. Time can also be represented implicitly as situations (as in the situation calculus) or worlds (as in classical temporal logic). In an explicit representation of time, the basic units may be points or intervals. In addition, time can be modelled consisting of discrete units or as a continuous range of values (for instance, real numbers).

In our approach we use a (discrete) linear model of time, with both point-based and interval-based representations depending on the type of use. For instance, we may use time points when we state that a certain property has a specified value at a certain point in time, while we use intervals as the base unit of time for events. However, a time interval is constructed as a pair of time points, so that time points are the primitive concept in our approach.

Introducing Fluents. The formal construction to generalise the concept of dynamic properties is to introduce a special class of objects, called *fluents*. They encode dynamic properties of objects. Fluents are used, for instance, in STRIPS planning [75], where they refer to an atom of information that may or may not hold in a state during the course of actions planned. In the blocks domain, for instance, one such atom may be $On(b_1, b_2)$ stating that box 1 is on top of box 2 (at the current time).

Instead of representing truth values, we use a slightly different notion of fluents in our approach, which can be summarised as follows: A fluent represents a property of an object in terms of a value which is an object. This object itself is static in time. However, it can be associated (related) to different values at different times. For instance, a train may have different location (values) at different times, but the property (like the name "location") stays the same. A fluent can be conceived as a place holder, a variable in a programming language, or a slot to store different values at different times. The slot itself remains unchanged.

A fluent f has at most one value $value_f(t)$ at each point in time[2].

$$value_f(t_1) \neq value_f(t_2) \rightarrow t_1 \neq t_2$$

If an interval-based representation were used, the situation would be more complex: In that case, the time intervals associated with two different values would need to be disjoint. In our model, changing the value of a fluent takes time. The time taken to change the value may depend on the values (for instance, the source and destination locations in a travel). A fluent stays constant between changes. This is modelled by a predicate *constant*/3 whose arguments are the fluent and the start and end time. This predicate is transitive. There are primitive (or atomic) changes and complex changes (that can be decomposed into primitive changes). During a change event the value of the fluent is undefined.

In the TRAINS domain, for instance, movable objects (engines, commodities, boxcars, etc.) have fluents called *location*. Each boxcar in addition has a fluent called *load*. In this domain, objects do not share fluents: each fluent is associated with exactly one object in the domain. In other domains, this might not be the case. There is no obvious argument against sharing a fluent

---

[2] A subscript notation is mainly used here for the sake of readability, formally, f is an argument like the others.

among multiple entities, but neither is there an apparent situation where this would be advantageous for typical entities. For instance, if an engine shared its location property with a boxcar, then they would be forced to be at the same location at any time. However, this would not make much sense, because one would not be able to model the attaching or detachment of cars. Thus, properties are only shared for the duration of certain events, such as travels.

Implementing Fluent Changes as Events. Apparently, the important aspect of fluents is the question of how changes to the value of a fluent are constrained, i.e. under which conditions changes are possible and if there are interactions with other entities. In particular, the underlying practical question is how to model fluent changes in order to make them work in a model generation approach.

The basic approach to represent information about a value change in a fluent f from value x_1 at time t_1 to some value x_2 at time t_2 is to introduce a relation like:

$$change_f(t_1, t_2)$$

As a remark on notation, in some circumstances, it makes sense (for efficiency reasons) to include the actual values in the representation, such as x_1, x_2 in $change_f(t_1, t_2, x_1, x_2)$, even though that information is redundant, given $value_f$. We make use of both representations.

As mentioned above, we want to represent change events as entities in the domain. However, it is not advantageous to directly use the $change/3$ relation[3] to infer these entities. For n value statements $O(n^2)$ changes would be inferred. The idea is to infer $O(n)$ ordered change events that explain how the property has changed its value and in which order.

In order to do so, one may distinguish between primitive and complex changes which consist of a sequence of primitive changes. An atomic change is one that cannot be divided into smaller changes. Event objects, from now on just "events", represent an atomic change of a fluent from an original to a new value. Complex changes are represented as entities in the domain.

Fluent change events have the following properties:

1. Each event has a unique starting and ending time, defining its duration.
2. Each event e is related to at least one fluent f, through a relation of the form $fluent(e, f)$.
3. Each event has unique old and new values it assigns to the fluent at its starting and ending time.
4. Two distinct events changing the same fluent cannot overlap in time. However, they (their temporal extensions, respectively) can meet, i.e. the value may hold only at one instant of time that is both the end of the first and the start of the second event.

An event may refer to more than one fluent. However, because of properties 1 and 2, all fluents are changed synchronously from the same old to the same

---

[3] The arity is 3 because f is used as an argument.

new value. For instance, in the TRAINS domain, a move event that changes the location of a product simultaneously changes the locations of the boxcar it is loaded into, as well as the engine that pulls the whole train. Perhaps it is not necessary to model these changes as one change event, but it seems to be an efficient and concise way to do so.

The next question to address is when an event is created in the sense that it needs to be represented as an entity. An event can be inferred whenever it is known that a fluent has two different values. The problem is that this again applies to $O(n^2)$ combinations in the case of $O(n)$ value statements. In order to prevent the creation of a quadratic number of events and to determine which of these combinations actually lead to the same change event, two auxiliary functions have to be defined: *next fluent change event* function $N_f(t)$ and *last fluent change event* $L_f(t)$. $N_f(t)$ refers to the next event changing fluent f on or after time t. Conversely, $L_f(t)$ yields the last event ending at time t or before. In order to formalise this, we postulate the following:

$$N_f(t) = e \rightarrow constant_f(t, start(e)).$$
$$L_f(t) = e \rightarrow constant_f(end(e), t).$$

If a fluent f has a value at one time instant t_1, and a value at a later time instant t_2, then either the fluent stays constant during the whole period between t_1 and t_2, or there is a (primitive or complex) change of f between these two time instants. If there is a change, there is a next and a last fluent change event at the borders of the time period. The two events may be identical (if and only if it is an atomic change.)

$$value_f(t_1) = x_1 \wedge value_f(t_2) = x_2 \wedge t_1 < t_2 \rightarrow$$
$$constant_f(t_1, t_2) \vee change_f(t_1, t_2, x_1, x_2).$$

If there is no change, then obviously the values have to be identical. Also, constant is a homogeneous predicate, i.e. in subinterval is also constant:

$$constant_f(t_1, t_2) \rightarrow value_f(t_1) = value_f(t_2)$$

$$constant_f(t_1, t_2) \wedge t_1 < t < t_2 \rightarrow constant_f(t_1, t) \wedge constant_f(t, t_2)$$

A change, on the other hand, implies the existence of the "next" and "last" change events, as introduced above:

$$change_f(t_1, t_2, x_1, x_2) \rightarrow$$
$$(\exists e \: . \: N_f(t_1) = e \wedge old(e) = x_1)$$
$$\wedge \: (\exists e \: . \: L_f(t_2) = e \wedge new(e) = x_2)$$

In this formalisation event entities are created only indirectly through the N and L functions. Notably, these only depend on one of the time instants.

Furthermore, a change must be either atomic or complex:

$$change_f(f, t_1, t_2) \rightarrow change1_f(t_1, t_2) \vee change2_f(t_1, t_2)$$

One complication here, concerning mostly the reasoner, is the fact that, in general, the theory cannot predict whether $change_1$ or $change_2$ applies. From the knowledge of two different value statements, it cannot be inferred if there isn't a third one between them. In the TRAINS domain, for instance, if it is known the engine E_1 is in Avon at t_1 and in the neighbouring city of Bath at t_2 (such that $t_1 < t_2$), it is logically not guaranteed or required that the respective move is a direct one (shortest path). And even if only one route were possible, E_1 may go to Bath, then somewhere else (e.g. Corning), and then back to Bath and t_2 could refer to this later instance. On the other hand, in special cases, the time constraints between the values may rule out a $change_2$. For instance, again in the example above, if there is less than 6 hours between t_1 and t_2 then only the direct route is possible.

Now both cases of changes are formalised individually. An atomic change is only one event:

$$change_{1f}(t_1, t_2) \rightarrow N_f(t_1) = L_f(t_2)$$

A complex change implies at least two events preceding each other, and thus they cannot be identical:

$$change_{2f}(t_1, t_2) \rightarrow event\_precedes(N_f(t_1), L_f(t_2))$$

It may be controversial whether or not all complex changes need to be represented as entities in the domain. Using natural language, one may refer to complex changes as entities. However, for the purpose of representation they are not necessary.

Furthermore, we use the following axioms to relate the definition of $fluent$ to those of N_f and F_f:

$$
\begin{aligned}
&fluent(e) = f \rightarrow \\
&\quad L_f(end(e)) = e \\
&\quad \wedge \; N_f(start(e)) = e \\
&\quad \wedge \; event\_precedes(e, N_f(end(e))) \\
&\quad \wedge \; event\_precedes(L_f(start(e)), e)
\end{aligned}
$$

Figure 4.2 illustrates these dependencies, given the change of a fluent's value between two instants of time t_1 and t_2. If a change can be inferred, for instance, because the values are not identical, then there is a next change event $N_f(t_1)$ after t_1 and a last event $L_f(t_2)$ before t_2. These two events may be identical (in the case of a direct change, cf. (a)) or preceding each other (cf. (b)).

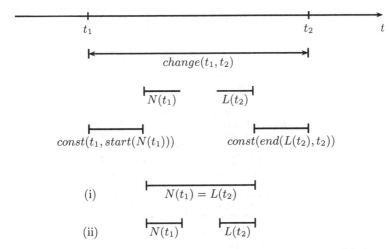

Fig. 4.2. The temporal dependencies in an inferred property change. $L(t_1)$ starts on or after t_1 and may be (i) identical to $L(t_2)$ which ends before or at t_2. Alternatively (ii), $N(t_1)$ may precede $L(t_2)$. The subscript is omitted for readability.

4.4 A Library of Reusable Domain Theories

This section provides an overview of our implemented domain models and their interrelations (cf. Figure 4.3). A library of reusable domain theories has been implemented. The concepts the domain theories introduce range from very abstract and generic ones, such as time instants or lists, to application-specific concepts, such as the TRAINS domain that will be described in the next section.

The library is organised as a collection of separate *modules*. When we speak of a domain theory, we mean the module of the same name and all dependencies included. For the use in the reasoner (cf. next chapter), a simple module system is provided to package module including all its dependencies into a single such domain theory. In particular, the term "application domain theory" denotes a domain theory which is complete in the sense that it may be used for reasoning about specific problem instances in that domain. By including or not including specific modules, the module system can also be used to enable or disable certain features of the library. Each module of the library may introduce classes (unary predicates), functions, as well as binary and other relations. In the following, the essential base modules and some domain theories will be described in more detail.

4.4.1 Basic Arithmetics

The Arithmetics module is used for representing a finite discrete numeric domain. It includes a less-than-equal and summation relation.

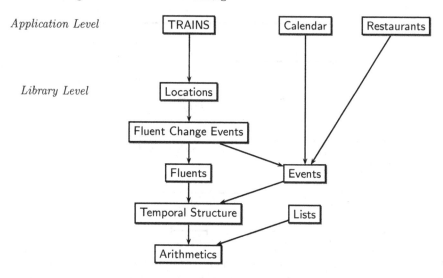

Fig. 4.3. The principal connections among the modules of our library of domain theories and the application domains.

It can be used, in particular, for representing time instants, i.e. we map time instants to integer numbers. For instance, in the basic calendar domain, as well as in the TRAINS domain, the numbers represent hours on a 24-hour scale. Thus, comparing time instants and calculating time differences can be accomplished.

The following predicates are defined:

number/1:
>   This predicate defines the basic class *number*. In our approach, this corresponds to a finite set of discrete values. Different application domains may use different sets of numbers.

leq/2, *lt*/2:
>   These predicates represent (complete) orderings on the *number* entities. They correspond to $\leq$ and $<$, respectively.

sum/3:
>   This relation defines the arithmetic sum of the first of the two arguments. It can also be used as a function *sum*/2.

In principle all tuples of the sum predicate can be listed for finite domains. In addition, the *leq*/2 relation can be derived from a successor function *succ*/2 defined on numbers. Furthermore, a strict ordering *lt*/2 (less than) can be defined, if required.

From a reasoning perspective, one has to keep in mind that for n elements of the *number* type, $O(n^2)$ ordering and *sum* statements will be derived. In view of that, the creation of new terms has to be concerned. As we will see later when the reasoning engine is discussed, even the presence of logically redundant clauses can influence the overall behaviour of the system. For instance, adding the clause

$$number(x_1) \wedge number(x_2) \rightarrow leq(x_1, x_2) \vee leq(x_2, x_1)$$

is logically redundant, since $leq/2$ is defined for all instances of $number$. However, in the reasoning process (which will be described in the next chapter), x_1 or x_2 may initially be uninstantiated, and then the presence of this clause may trigger a branch that would otherwise not be created. Therefore, this clause is factored out to a separate module that can be included by an application developer if desired.

In another module, a predicate $plus\_leq/3$ is axiomatised. This may be defined as

$$plus\_leq \equiv \{(x, y, z) \mid x, y, z \in number, x + y \leq z\}$$

The purpose is to be able to express that the difference between two numbers is at least y and to perform inferences of the form:

$$plus\_leq(x_1, 2, x_2) \wedge plus\_leq(x_2, 5, x_3) \rightarrow plus\_leq(x_1, 7, x_3)$$

This is useful, for instance, when the x_i represent time instants in order to reason about the combined duration of two events whose exact starting and ending times are not known.

Another general decision concerns whether or not to explicitly model finite ranges of integer numbers at all. An alternative option might be to implement the operations, like addition, directly in the reasoning engine. However, for the sake of consistency and also to keep our approach "conceptually scarce" we have decided to avoid these kinds of special behaviours.

4.4.2 Temporal Structure

The Temporal Structure module implements predicates for specifying time points and intervals. A discrete linear model of time is implemented. In a separate module, which is added for the domains we have considered, time points are identified with numbers. In that case this module is reuses concepts from the basic Arithmetics module. One of the main uses of both time points and time intervals from this module is to represent events and fluent values.

The module introduces the following classes of entities:

timepoint/1:
    The *timepoint/1* predicate specifies the base class of time point entities.
timeinterval/1:
    This class represents time intervals. It can be implemented on the basis of time points. However, this is not required.
timeinterval_empty/1:
    This subclass of *timeinterval* represents empty time intervals.

In addition, we define the following time point-based relations, corresponding to the *leq*/2 and *plus_leq*/3 predicate in the basic arithmetics module:

timepoint_precedes/2, *timepoint_strictly_precedes*/2:
:   These predicates define a temporal order on timepoints. *timepoint_strictly_precedes*/2 disallows identical timepoints.

timepoint_precedes_by_at_least/3:
:   This predicate is used for specifying an extended form of the previous relations. If numbers are used in order to refer to timepoints, *timepoint_precedes_by_at_least*(t_0, t_1, d) is logically equivalent to *timepoint_precedes*$(sum(t_0, d), t_1)$.

For intervals, we implement the temporal relations introduced by Allen and Ferguson [73] on the basis of the following relations:

timeinterval_precedes/2:
:   This predicate states that a time interval is before another. This relation is transitive.

timeinterval_meets/2:
:   *timeinterval_meets*(i_0, i_1) states that the time interval i_0 ends at the start of the time interval i_1.

timeinterval_disjoint/2:
:   This predicate states that time intervals i_0 and i_1 do not overlap. This relation is symmetric.

timeinterval_overspans/2, *timeinterval_during*/2:
:   The first predicate states that i_0 completely contains i_1. This relation is transitive. Its counter-part is the relation *timeinterval_during*/2.

In case of a point-based implementation of time intervals, we use the relations *timeinterval_start*/2, *timeinterval_end*/2 to represent the structure. A time interval is defined in terms of its starting and ending timepoint. The semantics of an interval is the right-open case, i.e. the starting timepoint may be part of the interval, the ending is not. Thus, a time interval with the same starting and ending time is empty.

4.4.3 Lists

The Lists module provides the definition of basic list structures. A *list* is an ordered sequence of objects. It is constructed out of *head* and *tail* properties. Tail lists can be shared. Like in a functional programming language, in the logic-based approach data structures cannot actually be modified, only new objects can be created from scratch or out of existing ones.

Lists are useful in many situations. For instance, in the TRAINS domain, they are used to represent the collection of transport cars attached to an engine. To this end, a number of operations needs to be logically defined, such as removing an element from a list.

This module defines the following classes and additional predicates:

list/1:

This predicate specialises into *empty_list*/1 and *non_empty_list*/1.

empty_list/1, *non_empty_list*/1:

These predicates defines the classes of empty and non-empty lists.

cons/3:

This is a constructor predicate that relates a non-empty list out of a head element and a tail list.

head/2, *tail*/2:

Corresponding to *cons*/3, these functional predicates may be used on a non-empty list to refer to its first element and its tail list, respectively.

remove/3, *insert*/3:

The predicate *remove*/3 implements a function returning a list with the first instance of a given element removed. In this definition, it must be present in the list. The counter-part of *remove*/3 is *insert*/3.

is_sublist/2:

This predicate defines a relation between lists where the first list can be constructed from the second by leaving out arbitrary elements.

is_filtered_list/3:

Similarly to *is_sublist*/2, this predicate defines a sublist relation between the first and the second list arguments. However, in this case a filter predicate has to be defined that determines which elements to remove.

append/3:

This predicate defines the relation between three lists l_1, l_2, l_3 such that l_3 is the result of the concatenation of l_1 and l_2. This relation is a function in its third argument. A recursive formulation of appending can be given as follows:

$$append(l_1, l_2, l_3) \land empty(l_1) \rightarrow l_2 = l_3.$$
$$append(l_1, l_2, l_3) \land non\_empty(l_1) \rightarrow cons(head(l_1), append(tail(l_1), l_2), l_3).$$

is_non_repeating_list/1:

This predicate is used for implementing sets on the basis of lists without sorting.

$$empty\_list(l) \rightarrow is\_non\_repeating\_list(l).$$
$$is\_non\_repeating\_list(l) \land non\_empty\_list(l) \rightarrow is\_non\_repeating\_list(tail(l)).$$
$$is\_non\_repeating\_list(l) \land non\_empty\_list(l) \rightarrow non\_member(head(l), tail(l)).$$

length/2:

This functional relation returns the number of elements in the list. Its definition is based on the *sum*/3 relation from the arithmetics module.

member/2:

A relation between the elements of a list and the list itself.

non_member/2:

In certain situations it may make sense to explicitly represent information about entities which are *not* members of a list.

A particular subclass of lists are sorted lists. These may be used, for instance, to approximate sets. For the purpose of modularity, declarations for sorted lists are factored out in a separate module. This module defines predicates useful for working with lists whose elements are sorted in a certain order. In particular, application domains may specify a limited form of comparator relations.

strictly_sorted_list/1:
:   cf. domain on sorted lists.

sort_less_than/2:
:   This predicate defines a (default) sort order on any entity. It should be defined for any entity that is an element of a default sorted list.

compared/3:
:   This predicate defines a comparison function. It takes two arguments and "returns" the comparison result as the third. It is an alternative to using *sort_less_than/2*.

compared/4:
:   This predicate defines a way to create different sort orders. The first argument is the "name" of the sort order. The remaining arguments take the same role as the arguments of *compared/3*.

is_sorted/1:
:   This predicate requires its argument (a list entity) to be sorted by the default order.

is_sorted_by/2:
:   This predicate functions in much the same way as *is_sorted/1*, but provides the possibility to specify the comparison function (actually, the "name" as defined with *compared/3*) to be used.

sort/2, sort_by/3:
:   These relations associate list with the sorted versions of it.

4.4.4 Events

Events are abstract domain entities that contain a temporal extension, characterised by a time interval consisting of a starting and ending time. The most important specialisation of this abstract class are so-called fluent change events which are described in Section 4.4.5. In our model, the time interval must be non-empty, i.e. the starting time has to strictly precede the ending time.

The following classes and relations are defined in this module:

event/1:
:   This is the base predicate for event entities.

time/2:
:   This relation associates the event with its temporal extension, a time interval. The time interval is required to be non-empty.

event_precedes/2, *event_overlaps*/2:
  In analogy to time points, events can be temporally related in different ways, such as preceding, meeting, or overlapping. These predicates can be derived from the respective predicates defined on the event's temporal extension (i.e. *timeinterval_precedes* etc.).

resource/2:
  For certain instances of events overlapping may be prohibited. For instance, fluent change events affecting the same fluent are not allowed to coincide. This is generalised to a *resource*/2 relation that excludes overlappings with events affecting the same resource or resources.

4.4.5 Fluent Change Events

The Fluent Change Events module provides a definition of events that model changes in fluent values. The essential properties of fluents and fluent change events have been introduced in Section 4.3 as part of our approach to modelling events and changes; this section may serve as a summary and reference.

Essentially, fluents are mappings from time instants to values. Fluents can be compared to variables in an imperative programming languages. Essentially, they are a way to refer to a slot in which different values can be stored at different times.

The classes and relations introduced in this module are described in the following:

fluent/1:
  This predicate defines the class of entities that can be used as fluents.

value/3:
  The relation *value*/3 can be viewed as a function from a pair of fluent entity and a timepoint (f, t) to a value v. Using time points as the base unit enables reasoning about value assignments that are only valid at one point in time.

constant/3:
  An atom of the form $constant_f(t_1, t_2)$ states that the fluent is constant in the indicated interval. This relation is transitive and homogeneous.

change/3:
  An atom of the form $change_f(t_1, t_2)$ indicates that the fluent f changes its value between time t_1 and t_2, i.e. some time in this interval. Since a value change takes time, this interval must be non-empty. This relation is transitive.

Fluent change events are a subclass of generic events, i.e. objects associated with a temporal extension. Thus, fluent change events inherit from the event type, the start, end, and duration properties. Fluent change events represent primitive (atomic) changes to fluent values, from an initial (old) value to a new one. These are represented by respective (functional) relations.

The essential characteristic of a fluent change event is that it is associated with one (or more) fluents. This is encoded in the relation $fluent/2$. The fluent (or fluents) have exactly one old and one new value which are to hold at the start or at the ending time, respectively. This implies that a fluent change event can affect more than one fluent. However, it has to change them in the same way, i.e. change them from the same initial value to the same final value. This mechanism is used, for instance, in the TRAINS domain for moving complete trains consisting of an engine, any attached transport cars, and potentially multiple packages of products, with only one fluent change event. Two (non-identical) fluent change events that affect the same fluent cannot overlap, i.e. one has to precede the other.

Fluent change events relate to the special predicates:

$N/3$, $L/3$:
   Fluent change events are also referred to in the N and L relations. These denote the next (respectively last) fluent change event after (respectively before) a certain time point.

4.4.6 Locations

The Locations module is used for representing high-level information about locations and movements. To this end, the distinction between mobile and immobile objects is introduced, and mobile entities are associated with a location fluent.

The Locations module introduces the following predicates:

$mobile/1$, $immobile/1$:
   These predicates define the disjoint classes of mobile and immobile entities.

$fluent\_location/2$:
   This predicate defines a function on mobile entities that associates the entity with a fluent. This fluent represents the location of the entity at a given time instant. The value of the fluent is required to be an immobile entity.

$route/1$:
   A route is an ordered sequence of multiple locations. A route may be composed of individual route segment entities.

$route\_duration/2$:
   This relation associated a route or route segment to its duration.

Travel or logistics domain applications can be based on the Locations module. They would typically provide a database of locations and route segment instances. Additional extensions of this module may include

- a classification of different means of transportation, for instance, car (private, rental, taxi), bus, train, plane, ship, walking. Each of which may include particular restrictions and the domain may also model switching between these means of transportation.

- models for the ticket reservation and ordering process.
- models for delays in route segments, caused by, for instance, traffic jams, blocked roads, etc.

4.4.7 Relational Databases

Rather than a specific module, this section describes a general approach to represent the contents of a basic relational database. In many cases a domain theory includes information about entities whose properties can be represented in a table, such that each row represents one entity, and each column represents a particular property. This kind of data can be straightforwardly represented in the domain theory. However, this representation can be roughly characterised as low-level statements of the form "table A, row B, column C, value D", i.e. there is not automatic high-level interpretation as to what the meaning of a row in a certain table is. Nevertheless, such an interpretation can be accomplished in a domain theory with logical mapping rules.

Let us assume a table tab with columns $col_1, \ldots, col_n$, m rows, and corresponding values $val_{1,1}, \ldots, val_{m,n}$. We assume the table's first column is a primary key and the columns forbid NULL values. We later generalise to avoid the latter restriction. The table can be represented by the following theory:

$$is\_table\_row(tab, val_{1,1}).$$

$$\vdots$$

$$is\_table\_row(tab, val_{m,1}).$$

$$is\_table\_row(tab, x) \rightarrow x = val_{1,1} \vee \ldots \vee x = val_{m,1} \qquad (4.1)$$

$$has\_value(tab, val_{1,1}, col_1, y) \rightarrow y = val_{1,1}.$$

$$\vdots$$

$$has\_value(tab, val_{m,1}, col_n, y) \rightarrow y = val_{m,n}.$$

These axioms list the contents of the table. Note that it is not necessary to represent the declared column of the table, since they are implied by the values. Axiom 4.1 is used to close the table, i.e. to state the list of rows is complete. In a reasoning process, when an atom $is\_table\_row(tab, x)$ is processed, this will lead to the creation of a large number of branches. It is thus sometimes advantageous to derive specialised versions of axiom 4.1 in order to avoid this behaviour in the following way:

$$is\_table\_row(tab, x) \wedge has\_value(tab, x, col_j, val_j) \ldots \rightarrow$$
$$x = val'_{1,1} \vee \ldots \vee x = val'_{m,1}$$

The idea is to use additional information about x (i.e. that x has a certain known value in a certain column) that may be available in order to restrict the possible choices of x. Ideally, there is only one candidate x remaining, and thus the disjunction can be avoided.

Foreign keys are an important tool in relational databases to ensure the referential integrity of the database. They are SQL instructions of the following kind:

```
ALTER TABLE tab
ADD FOREIGN KEY key1  (colJ,...)
                  REFERENCES  tab2 (colK,...)
```

A simple foreign key (only one column that maps to the primary key of the referenced table) can be represented by constraints of the following form:

$$is\_table\_row(tab, x) \land has\_value(tab, x, col_j, y_j) \rightarrow is\_table\_row(tab_2, y_j)$$

SQL provides the possibility to construct *views*. Views can be seen as virtual tables that combine information from different existing tables (or other views.) Views are defined through a **SELECT** statement that references the data sources. Since in our case we are only modelling a static state of the database, i.e. its contents at one point in time, we can handle the view in the same manner as a regular table in order to represent its content.

Above we assumed non-null values in the table content, here we describe how to avoid this restriction. The NULL value is used in a database table to indicate that a certain value of an object (row) is not applicable (possibly depending on other values in the same row) or that it is not known. This incurs the need for different ways to represent columns which allow NULL values. The NULL value can be represented as a special value, that would be checked with a predicate $is\_null/1$. Alternatively, for each nullable column col, a new *virtual column* col' of type Boolean is introduced, indicating if the value in column col is null. In this case, the value in the original column is not to be used. In case the NULL value represents an unknown but applicable value, it might make sense to represent it by an uninstantiated Skolem term. However, this may pose a problem when additional constraints are inferred which cannot be represented in the database. The two approaches differ in whether they allow a $has\_value(tab, x, col, y)$ atom in the case of a NULL value or not.

The model as illustrated so far deals with the representation of a database state, and is able to ensure that the database's content is in compliance with various constraints. It also provides means to efficiently access the database content, for instance, through specialised inference schemata that infer row ID values based on certain column values. However, the model does not address the question what would constitute a valid *database modification*, and through which database modifications a transition from one database state to the other could be achieved. Essentially, reasoning about database modifications can be seen as a *meta-domain* over the database domain. The difficulties concerning reasoning about database modifications arise from the fact that different database states (at different times) have to be represented. One could model the dynamic content of a database table as a list-based fluent. An ontology of possible changes (insertion, deletion, update events) could then be defined.

4.4.8 Application Domains

This section describes some of the application domains that have been implemented using the library of domain theories described in the previous section. Since these domains are modelled on top the library of domain theories, they can be considered integrated on a basic level, i.e. they are using fluent change events and the same temporal model. In order to achieve a tighter integration of the domains, for each combination of domains, connecting axioms can be introduced. Also, the distinction between the application domains and the modules of the library is a rather loose one. One domain, for instance, may rely upon another as a module. One convention for distinguishing library modules from application domains is that application domains focus on the concrete inventory of entities they typically need to include to be useful, while modules focus on the conceptual level.

Calendar

The calendar application domain defines appointments which are derived from events, with the additional requirement that they also require a location property. Appointments, as events, can additionally be associated with resource objects representing entities, such as meeting rooms, equipment devices, persons, etc. which can only be assigned exclusively to one appointment at any time. That is, the assignment could also be expressed as a fluent of the resource with the appointment being the value. In addition, the resources need to be physically present for the duration of the appointment. The calendar domain defines the state of a calendar in terms of a set of appointments. This model is static, i.e. no changes to the calendar are modelled.

A potential problem in the calendar domain is the fact that a literally infinite number of entities exist in this domain. But since we are interested in finite models, we need to be able to extract relevant ranges of dates and an adequate level of granularity. For instance, if days or weeks are used as the base unit of an application using the calendar domain, it would not make sense to introduce entities for every minute in the range of interest.

The possible extension of a meta-calendar domain may define consistent operations on a calendar and relate the operations to the states of the calendar at different time instants.

The following operations on a calendar would need to be defined:

- creation, i.e. the scheduling of an appointment,

- modification, i.e. concerning the time, location, participants, or resources properties,
- deletion, i.e. cancelling an appointment.

The consistency of operations on a calendar involves at least the restriction that an appointment can only be modified before it takes place, not after that. There may be stricter additional time limits.

In order to model the states of a calendar explicitly, lists of calendar entities can be used. These can be used as values for a respective fluent. The operations on the calendar can be modelled as insertions, deletions, and substitutions on this list. The calendar operations can be implemented as fluent change events. The additional restrictions may be placed on the old and new values. In order to prohibit changes to an event once it has begun, the ending time of the fluent change event can be constrained to precede the starting time of the appointment object to be modified.

TRAINS

As outlined above, the TRAINS domain is based on the modelling of fluents and fluent change events. Fluents are used for implementing different dynamic properties of objects in the domains, such as an object's location or its load (in the case of transporters, for instance). Time is modelled as a finite sequence of discrete time points. In the TRAINS domain the values are restricted to (0-24), representing the 24 hours period of one day. This implies that time durations are multiples of one hour and all information encoded refers to one day only. The domain also contains declarations for the initial state of the TRAINS world (time 0). The TRAINS domain is described in more detail in Section 4.5.

Restaurants

The Restaurants domain defines basic information about restaurants, such as the cuisine or cuisines a restaurant offers, its location, name, and price range etc. These properties are typically used when selecting restaurants. This application domain includes data about specific entities. The data have been obtained from the Helmut project [85] and translated from their XML-structured source into a clausal theory. They describe about 100 restaurants in and around Ulm.

On top of this raw data the possible cuisine and location values have been generalised to hierarchical concepts, i.e. forming a simple ontology. These generalised concepts can be taken advantage of by a user who wishes to state some general requirements or preferences before committing to definite choice. For instance, the top-level concepts in the cuisine ontology contain concepts such as "Asian cuisine", "European cuisine", or "fast-food". Likewise, the location information is organised in a (simple) geographical hierarchy, that first distinguishes between the city centre and the outskirts and on the second level between the districts. The price range options are modelled as discrete classes from budget to expensive. Various other features, such as smoking/non-smoking information or information about the general type of restaurant (pizza delivery service vs. luxury restaurant) have been neglected for the time being but could be integrated at a later stage.

4.5 Modelling the TRAINS Domain

This section discusses an application of our modelling approach to the TRAINS domain. It serves as a showcase for the techniques presented so far, in particular, the fluent-based modelling of dynamic properties of entities. Apart from the domain-specific information about the TRAINS domain that will follow, it is important to note that this domain is interesting because important aspects can be generalised to other domains.

The TRAINS domain (cf. Figure 4.4) was originally devised for experiments conducted in order to collect natural language dialogues between two human interactors [50]. The rationale behind these experiments was to find out what an automatic system should be like in order to behave as natural as possible. The dialogues were performed by two persons connected through a telephone setup, so there was no visual communication channel. The two interactors were assigned different roles: one person acted as the *system*, i.e. as an expert in the domain, but without own concrete goals. The other was to enact a *user* who wished to solve a concrete problem in the logistics domain. In order to emphasise the distinct roles and to stimulate communication between the two interactors, the user was given (a map with) a substantially smaller amount of information concerning where resources were located and how long actions, such as moves, were to take.

Although the TRAINS domain was intentionally conceived as a toy world, it is far from trivial as a problem-solving domain. There are a large number of interdependencies between actions and it is, in many cases, non-trivial to prove that one has indeed found the simplest solution to a given problem. For instance, if a certain amount orange juice is to be transported to one city, it has to be produced first, which in turn requires to transport oranges in time. In addition, transporting commodities requires transport cars which may be picked up in different locations.

An additional factor that makes the TRAINS domain challenging for an automatic reasoner is the fact that there is a very large number of choices concerning resources for each action. Many choices and parameter configurations are logically consistent, but a human would probably not consider them (at least, right away). For instance, if a transport goes via different stops it is in principle possible to change the engine in each city without further costs. However, humans would tend to keep the same engine. On the other hand, problems might be constructed that would require the changing of engines in order to find the minimal solution.

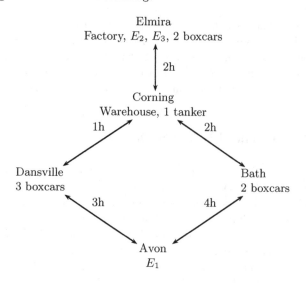

Fig. 4.4. The TRAINS logistics domain [50].

Static Model. The TRAINS domain consists of the following concepts (which can be represented in a conventional class hierarchy, cf. Figure 4.5):

- Cities and tracks connecting cities: There are five cities and five tracks, with different durations.
- Cars, which are either engines or transport cars. There are three engines and seven transport cars.
- Transport cars are either box cars for solid products, or tankers that can load orange juice.
- Packages of products, which are either oranges or bananas.
- For each commodity, there is a warehouse which initially stores the complete supply of packages. The warehouses for oranges and bananas are in Corning and Avon, respectively. In addition, there is an orange juice factory in Elmira which can produce orange juice from raw fruits.

Entities of these classes can be listed as it would be done in a relational database. For instance, the cities are represented by the following domain theory fragment:

$$city(\text{``Avon''}).$$
$$city(\text{``Bath''}).$$
$$city(\text{``Corning''}).$$
$$city(\text{``Dansville''}).$$
$$city(\text{``Elmira''}).$$
$$city(x) \rightarrow$$
$$\quad x = \text{``Avon''} \lor x = \text{``Bath''} \lor$$
$$\quad x = \text{``Corning''} \lor x = \text{``Dansville''} \lor x = \text{``Elmira''}.$$

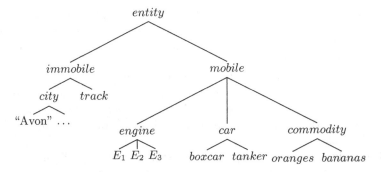

Fig. 4.5. The classes of entities used in the TRAINS domain. For the case of *city* and *engine*, corresponding entities are shown.

Here, the last clause states that these entities are in fact all of the type "city".

The *tracks*/4 predicate combines different pieces of information, including the distance (in hours), and relates them to the first argument, the track name (or ID).

$$track(\text{"trAB"}, \text{"Avon"}, \text{"Bath"}, 4).$$
$$track(\text{"trAD"}, \text{"Avon"}, \text{"Dansville"}, 3).$$
$$\vdots$$

These individual properties are dually encoded in binary functional predicates *origin*/2, *destination*/2, and *duration*/2, which take the track ID as their first argument:

$$track(x, f, t, d) \rightarrow track(x).$$
$$track(x, f, t, d) \rightarrow origin(x, f).$$
$$track(x, f, t, d) \rightarrow destination(x, t).$$
$$track(x, f, t, d) \rightarrow duration(x, d).$$

Note that in our approach the tracks are modelled as directed routes, i.e. there is one entity for each direction. Another predicate *same_track*/2 is required for encoding the knowledge which tracks are identical in the undirected sense.

The declarations so far can be seen as the translation of a relational database. While logically sufficient, this representation can be optimised using logically redundant axioms that are more efficient in some respects. For instance,

$$track(tr, \text{"Avon"}, dest, dur) \rightarrow tr = \text{"trAD"} \lor tr = \text{"trAB"}.$$
$$track(tr, \text{"Bath"}, dest, dur) \rightarrow tr = \text{"trBC"} \lor tr = \text{"trBA"}.$$
$$\vdots$$

These clauses encode "short cut" inferences that apply when the origin of the track is known. Similar short cuts can be derived automatically for each property of the tracks class, and in fact for any similar property of an object.

Similarly to the city and track class and entities, engines, boxcars, and tankers can be listed. Commodities (which subsumes oranges and bananas) are modelled as "packages" or "units" of products, e.g. one load of oranges. The state of a commodity (i.e. orange juice) is modelled using a fluent $fstate$, so the actual entity remains the same if one load of oranges is processed into one load of juice.

Dynamic Model. Apart from the static knowledge discussed so far, a large amount of the modelling is concerned with the dynamic aspects of the domain. In particular, an initial state is defined which represents the knowledge about the domain entities at a defined initial time 0. All movable entities have an initial location which is listed in the following way:

$$at(E_1, 0, \text{"Avon"}).$$
$$at(E_2, 0, \text{"Elmira"}).$$
$$at(E_3, 0, \text{"Elmira"}).$$
$$at(\text{"b1"}, 0, \text{"Dansville"}).$$
$$\vdots$$
$$oranges(o) \to at(o, 0, \text{"Corning"}).$$
$$bananas(o) \to at(o, 0, \text{"Avon"}).$$

The following dynamic properties need to be modelled:

- the location of mobile entities, such as cars
- the load of cars that can transport commodities
- the list of cars attached to an engine (i.e. forming a trains)
- the state of processing of commodities
- the current train of a track: the train that is on the track at a certain time, if any.
- the number or list of available entities of a certain type at a certain location and time.

The following informally described main rules govern the domain and their formalisation will be the task of the logic-based domain model:

1. Tracks connect locations (cities). There is at most one track between any two cities, and any connection can only be used by at most one train at any time, independently of direction. Cities can store any amount of commodities and cars.
2. Moving an entity (changing its location of an engine, car, or commodity) takes time depending on the origin and destination. There are elementary moves along one track from an origin to a destination, as well as complex moves that are composed of elementary moves. Elementary moves may take more than one unit of time, but cannot be interrupted or aborted.

If a train leaves from an origin, it is known to arrive at the destination exactly after the travel duration specified for the track.

3. Transport cars can only move if they are attached to an engine. Attaching cars to an engine does not take time, but obviously the engine must pick up the car at the car's location. There is a limit of three on the number of loaded cars attached to one engine. Cars can be attached to at most one car at any time. Cars can also be dropped at a city. Similarly, commodities only moved if they are loaded in a transport car. However, the loading and unloading of a commodity takes one hour of time. Loading can only take place if the product and the transport car are in the same city location at the same time for at least one hour (i.e. not during moves).

4. The processing of oranges into orange juice constitutes an event that combines the simultaneous unloading of the (solid) oranges from a boxcar and the filling of the resulting juice into a tanker car. Only one load of oranges can be processed at any time at the factory.

Our abstract approach based on fluent changes models how these move statements translate into actual events. An event corresponds to an elementary move of an object along a track. The requirement that commodities and transport cars can only move in parallel with an associated engine can be formulated quite concisely. In fact, the movement a boxcar and movement of its engine constitute aspects of the one and the same event.

In order to model attaching a boxcar to an engine, each engine e has a dynamic list-valued property $cars(e)$ denoting the list of currently attached cars. Attachments or detachments are equivalent to events changing this list value. The restrictions that the list can only change when the engine is in a city and that it can only grow when a (currently unattached) boxcar is also there, can be expressed by relating different fluents:

$$change_1(cars(e), t_1, t_2) \rightarrow constant(loc(e), t_1, t_2).$$
$$change_1(cars(e), t_1, t_2) \wedge member(c, value(cars(e), t)) \rightarrow$$
$$constant(loc(c), t_1, t_2, value(loc(e), t_1)).$$

The modelling of loading and unloading actions is similar to the attachment of boxcars. However, loading and unloading take one hour of domain time and during that time, the boxcar has to be located in the city where the event takes place.

As an optimisation regarding the possible move actions, minimal travel times are stated for travels between any two locations (direct or indirect):

$$min\_travel\_time(\text{"Avon"}, \text{"Bath"}, 4).$$
$$min\_travel\_time(\text{"Avon"}, \text{"Corning"}, 4).$$
$$min\_travel\_time(\text{"Avon"}, \text{"Dansville"}, 3).$$
$$min\_travel\_time(\text{"Avon"}, \text{"Elmira"}, 6).$$
$$\vdots$$
$$min\_travel\_time(x, y, z) \rightarrow min\_travel\_time(y, x, z).$$

In addition, the functional relation det-track($orig, dest, dur, track$) is meant to denote that in a journey from $orig$ to $dest$ given at most dur hours of time, the only choice for the next track is $track$. Such functional inferences can be handled much more efficiently in the reasoning process.

$$det\_track(\text{``Avon''}, \text{``Avon''}, 6, \text{``trAD''}).$$
$$det\_track(\text{``Avon''}, \text{``Bath''}, 5, \text{``trAB''}).$$
$$det\_track(\text{``Avon''}, \text{``Corning''}, 4, \text{``trAD''}).$$
$$\vdots$$

In order to support flexible expression of constraints, the knowledge within the domain needs to be maintained (to some extent redundantly) in different representations. For instance, for each movable object (such as commodities or engines) the location at a certain time instant is represented explicitly, and this is useful and necessary for reasoning. For instance, one would like to know where engine E_1 is at 7 o'clock in the plan. However, in many situations one does not care about exactly which entities are at a certain location, but wants to express how many of their kind are at a location. Consider, for instance, the utterance "There are 3 boxcars at Dansville". So, to enable these kinds of expressions, list-valued predicates have to be introduced. Then, statements about the lengths of these lists may be expressed.

$$cars\_available\_at(loc, t, cs) \rightarrow city(loc).$$
$$cars\_available\_at(loc, t, cs) \rightarrow timepoint(t).$$
$$cars\_available\_at(loc, t, cs) \rightarrow list(cs).$$
$$cars\_available\_at(loc, t, cs) \wedge member(c, cs) \rightarrow car(c).$$

Using these definitions, the utterance above might be expressed as:

$$boxcars\_available\_at(loc, t, cs) \wedge length(cs, 3).$$

This essentially states what we can infer when we know that a certain list of product entities is available at a certain city at some time t. The other direction, inferring the exact list of entities from other facts is more difficult. We have to view the predicate $products\_available\_at/3$ as a fluent and reason about how changes can happen to the list of entities available at a location. We will find that changes to this list are related to move events of these entities. Whenever a movable entity arrives or departs, the city's list has to be adapted. On the other hand, if the membership of an entity does not change at some time t, it cannot arrive or depart at that time.

4.6 Discussion

In this chapter, we have described our approach to the modelling of different application domains for the use in Spoken Language Dialogue Systems. The description focused on two main topics. Firstly, a formalisation for the representation of actions and events has been developed on the basis of existing approaches. It was adapted with special concern for using it with a model generation reasoning engine. Secondly, a library of domain theories has been developed that is to serve as the basis of modelling applications. As an example, the modelling of the TRAINS domain has been described in detail. Modelling on the basis of this library provides a way to integrate application domains.

From this perspective, to the general question whether or not First-Order Logic is an adequate tool for knowledge representation we can at least give the answer that it is for the domains we have considered so far. Some researchers, e.g. [54], argue in favour of a more expressive logic with closer resemblance to natural language. However, from our point of view, this does not seem to be as necessary for modelling many application domains. The same can be said about the problems of FOL in representing epistemic knowledge: Planning to obtain some knowledge is difficult to model. In principle, the system would require epistemic knowledge, i.e. knowledge about itself and what it can infer.

This is related to some interesting questions concerning the representation of what we called meta-domains: When certain states of affairs are represented on the domain level, for instance, a database state or a travel itinerary, is it useful for representing a respective meta-domain, i.e. one that models modifications to the state of affairs, and if so, how can this be done? Or should this be modelled the problem-solving level? There are arguments both in favour or against modelling meta-domains on the domain level. On the one hand, modelling can be done using the presented approach and in some applications it makes sense to consider modifications part of the domain. On the other hand, the modelling is much more complex and the models are substantially larger if this is done. In addition, some questions typically associated with the problem-solving level cannot be handled in the meta-domain approach. For instance, the problem-solving level has a kind of introspection capability concerning the domain level. This is to say, the problem-solving level can inspect the domain level and detect what kinds of ambiguities exist, for instance, and initiate specific problem-solving actions for this situation. However, this cannot be modelled in the meta-domain approach, because such conditions require a real meta level of representation. Otherwise, an ambiguity on the domain-level is also an ambiguity on the meta domain.

In summary, we can say that we view our logic-based domain modelling together with the reasoning engine described in the next chapter, as a kind of programming system in the sense of pure Prolog [86]. As such, the development of domain theories is usually interleaved with testing it on specific problem instances and integrates these results. In particular, some relations are formalised with the typical use cases in mind.

5

Interactive Model Generation

5.1 Introduction

In this chapter, we describe the design and implementation of our Common Interactive Domain-level Reasoning Engine (*CIDRE*). It is used as a tool to "breathe life" into our logic-based domain modelling discussed in Chapter 4 in order to take advantage of these models for concrete instances of domain tasks.

We start by outlining the related component architecture of its usage environment. CIDRE is conceived as one module in a spoken language dialogue system. In particular, the reasoning engine is to be used as an interactive component that mediates between the dialogue manager (acting on behalf of the user) and the domain applications that provide domain-specific functionalities and services. Building on that we discuss some of the advantages and potential drawbacks of using finite model generation as a reasoning procedure. Next, the actual implementation of the reasoning engine and the algorithms used are described. Several issues concerning our implementation, such as the data structures used, are discussed. A distinctive feature of our engine is the *interactive protocol* it provides to enable the use of the engine as an interactive component, instead of an autonomous black box. We discuss the design and implementation of that protocol. Finally, in Section 5.5 we describe our approach to testing the reasoning engine as part of the development cycle and present related results. These provide more insights into the workings of the engine when applied in a concrete domain-specific setting.

For our purposes, finite model generation [78] is a deductive reasoning procedure that, given a theory, or set of first-order clauses, as input on return provides consistent finite models, or, alternatively, a proof of the inconsistency of the theory. We distinguish between *axioms* on the one hand which are first-order logic formulas containing variables and, on the other hand, ground units, called *facts*. A model returned by the model generation procedure can be viewed as an extension of the input theory with a set of facts such that each axiom of the theory is met.

CIDRE implements a tableau-based deduction algorithm enhanced with novel features for an interactive usage. In contrast to some existing (especially non-tableau-based) implementations of finite model generation [77] which focus on enumerating small models fast, our procedure attempts to cope with larger models in an interactive fashion. It is designed to work as a module that may be synchronised with a dialogue manager's information state and contributes information, and thus is not required to find complete solutions autonomously. One of the most important features for this kind of interactive functionality is the reasoner's capability to store a trace for each proof step that is taken. This enables, firstly, to trace inferences back to their original hypotheses, which is necessary for explanation and conflict resolution in dialogue. In addition, it enables the reasoner to keep track of inconsistent sets of hypotheses, so-called *nogoods*. Nogoods are reused in the search process and may potentially cut off a large subtrees of the search space.

In terms of architecture of the usage environment (cf. Figure 1.1), the crucial difference to a conventional dialogue systems is that the dialogue manager does not directly communicate with different domain applications. Instead, the reasoning engine can be seen as a mediator between the dialogue manager and the different domain applications. All components in the architecture operate asynchronously, such that the dialogue manager may be occupied with producing output or with interpreting the next user utterance, while the reasoner is still processing input based on the previous dialogue state. In addition, the processing performed by the reasoning engine is incremental in the sense that it is able to incorporate new information from the dialogue manager without abandoning all of its previous work. Both the dialogue manager and the domain applications access the *proof database* that is created by the reasoning engine to store the proofs it generates. The components can query the proof database in order to obtain new inferences from the reasoning engine. The new information can also be filtered as needed by the querying component. The components can create new tasks to be processed by the engine. Within a reasoning task, the domain applications can create proof structures in order to implement domain-specific reasoning functionality. For instance, a travel planning application may create a proof concerning the duration of a trip as soon as the origin and destination are given in a form understood by the application. The domain applications have to make sure that the proofs they create are sound, i.e. all preconditions that are used in the proof (which are available as separate proof structures) have to be referenced correctly. This will be described in more detail in Section 5.4.

5.2 Model Generation as a Reasoning Procedure

Advantages. It is worthwhile to discuss some of the advantages and potential disadvantages of using finite model generation as an inference procedure: First of all, model generation reasoning is based on the notion of satisfiability (i.e. consistency) of the given input. Apparently, it is a desirable feature of a reasoning engine to detect inconsistencies in the input assumptions in order to be able to resolve them. In addition, the models found by a model generation procedure are structures that are fairly easy to interpret [68]: They consist solely of entities in the domain and relations which are enumerated as sets of tuples over these terms. Thus, for instance, very efficient access to these relations is possible by simple look-up mechanisms.

Furthermore, some model generation approaches are relatively intuitive inference techniques (cf. PUHR method, Section 3.6). In these methods, only two kinds of proof steps are used. And the one that creates new information can be seen as straightforward application of the rules in the input theory with the existing knowledge to fill the rules' preconditions. The existing knowledge consists of ground facts and so does the consequence. The other proof step is used to split the tableau being built for alternative hypotheses. Thus, the inference technique should also be well suited for explanation and negotiation purposes.

Also, in the case of an underspecified input (rather than an overconstrained situation) a suitable model generation procedure, such as the PUHR method, can list all alternative extensions of the input in order to find a solution. This means that if the method comes up with a solution, it also provides information about which additional assumptions have been introduced and thus where alternatives might be possible. To a user, this will constitute valuable information and a basis for additional decisions.

Disadvantages. As a reasoning procedure that operates on full first-order logic, finite model generation has the implied theoretical computational limitations. It is only semi-decidable in general to determine whether or not a given input theory is satisfiable (i.e. whether there exists a logical model that satisfies it). It is also only semi-decidable to determine if a *finite* model exists. Some theories may only be satisfied with infinite models, which does not seem to be helpful for practical systems since the user will be interested in solving his tasks in finite time. If the input theory is indeed finitely satisfiable (i.e. it has a finite model), a minimal model can be found procedurally (cf. Section 3.6).

Besides, there is the general issue of exhaustive reasoning or "missing goal-directedness" of the reasoning procedure, which sometimes is an advantage and might may become a problem in other situations. This means that the reasoning engine does not work in a strictly goal-directed fashion in which it tries to proof a certain goal statement (as Prolog-style prover would do) or tries to build a plan for a certain goal state to be achieved (as a planner would do). Such systems decompose an attempted proof (goal) into smaller subgoals and recursively solve these. Therefore, once all subgoals have been proved, the overall proof is finished. The same applies to (some, for instance, hierarchical decomposition) planning systems, which also decompose a complex planning task into simpler tasks that are solved recursively. Instead, in our approach the model generation procedure is using all its available information. This is necessary because as any piece of information could lead to an inconsistency. Thus, it is required to check all combinations of assumptions.

Other potential problems concern the linguistic presentation of produced inference steps and models. For instance, sometimes the modelling may employ auxiliary predicates that have a rather technical meaning, but are not adequate for a direct translation into natural language. It is therefore difficult to verbalise statements containing these predicates. However, this problem may be more due to the kind of modelling applied, rather than the inference process.

Interaction. We argue that model generation has properties that make it an attractive basis for an inference engine in an SLDS context. The main problem is the incompleteness of the procedure as an autonomous system (i.e. it may fail to find an existing model). However, this problem is addressed using the reasoning engine as an interactive component. Thus, instead of relying on an autonomous prover, a human interactor can prevent the reasoner from descending into an infinite recursion. In particular, the interactive protocol that will be described in Section 5.4 will enable extensive external access to the reasoning engine while the engine is processing given tasks. The interactive protocol enables the reasoner to provide "real time" information (such as inferences) and to receive information. This can be used, for instance, to identify decisions the reasoning engine has taken (e.g. branching) in order to negotiate them with the user. Thus, branches may be discussed with the user and may be confirmed or dis-confirmed. In the latter case, the current search space will be pruned and the reasoning will continue with an alternative to the dis-confirmed assumption. In addition, inferences already generated may be discussed with the user. Thus, even if the reasoning process has not yet finished, certain portions of a solution can already be negotiated.

5.3 The Core Reasoning Process

This section presents the core reasoning process that is used to generate inferences on the basis of a given *input specification* consisting of a base theory and additional (task) information. We introduce the necessary representations, most notably the reasoning state data structure before describing the main iterative function in Section 5.3.3. This function performs one defined reasoning step on the basis of the reasoning state. The core reasoning process consists of an iteration of this function until it reaches a fixed point. One of the advantages of implementing the reasoner using an iterative function is to enable interactivity. This aspect will be covered in more detail in Section 5.4.

The state of the reasoner at run time primarily consists a tree structure of context nodes (the tableau). Each context represents a set of hypotheses (i.e. internally assumed propositions). The initial state contains only the base context that has no hypotheses. In each iteration, the reasoner determines one context to be refined. Each context maintains an active list, where items (proofs) that need to be processed are stored. If the chosen context has no items on its active list, or *agenda*, it can be reported as a solution. Otherwise, subcontexts are spawned for each alternative in the item. If there the set of alternatives is empty, an inconsistency has been detected and the context is closed. Otherwise, the subcontexts are added to the tableau structure as subnodes of the current context with one alternative hypothesis added to each of them. The hypothesis is combined with all axioms that match, resulting in new items of the context's agenda. In the following, we won't use the term context. Instead, the reasoning state data structure will assume this role.

The input to the reasoning engine CIDRE is a theory in the format used in Chapter 4 for modelling, i.e. a set of rules in clause form. Each rule is a disjunction of literals, where the negated literals can be viewed as preconditions and positive ones as postconditions. The set of preconditions is called the rule's body, and the postconditions are the head. As a preprocessing step, empty bodies can be replaced by single special atom $\top$ (true), whereas empty heads can be replaced by $\bot$ (false). This is used by the reasoning engine to detect the clauses to start with (those that have only $\top$ as a precondition) and to detect inconsistencies (once $\bot$ has been derived). The clauses are range-restricted, i.e. each variable appearing in the head must also appear in the body. This guarantees that when the rule body is matched against a set of facts, all variables in the head are bound to terms, and thus the head then consists of a disjunction of ground facts.

5.3.1 Data Structures

This section presents details concerning our implementation of the main data structures. These are used to represent the reasoner input and the reasoner's inferences. The internal reasoner state is described in the following section. The core data structures are *terms, atoms, clauses, proofs*. Except for the proofs, the definition of the data structure closely mirrors the definition of its logical counterpart (cf. Section 3.2). Another complex data structure, representing the reasoner state, is discussed in the next paragraph.

Terms. Terms are the most basic components and are used in the more complex data structures. Terms can be either variables or ground. Variables only appear in rules read from the input theory. In other situations, only ground terms will occur. Ground terms can have arguments, i.e. they can be of the form $f(t_1, \ldots, t_n)$ where the t_i are a finite list of subterms. An order $<_T$ is introduced on ground terms by comparing the following properties in this order: the depth of the term, the function symbol, the number of function arguments, the actual argument terms. If all of these properties match, the terms are equal. The ordering $<_T$ is used for efficient access to terms and other structures.

Atoms. Atoms consist of a predicate and a list of terms the predicate applies to. Literals are negated atoms. The term order $<_T$ induces an order $<_A$ on atoms in the following way: Given two atoms, if the predicate names or arities differ, they decide which one is less. Otherwise, the lists of argument terms are compared to each other. The atom order $<_A$ plays a role in the selection of atoms from the head of a clause, for instance.

Clauses. A clause is a list (to be interpreted as a disjunction) of literals. However, here a representation in implication form is implemented, i.e. the positive and negative literals are grouped into head and body sets, respectively. The main use of the clause structure is to represent the input specification that the reasoner operates on.

Proofs. A proof structure records an inference that is based on previous proofs, i.e. a proof can be viewed as a directed acyclic graph where the nodes are labelled with proof steps. Proof structures are built when clauses are used to derive new facts from existing ones. Given that the clauses are range-restricted, the matching of the preconditions guarantees that the postconditions are ground. Given a proof structure $\mathcal{P}$, we denote by $pre(\mathcal{P})$ the set of proofs used as preconditions in the matching. $post(\mathcal{P}) = atom(\mathcal{P}) : alts(\mathcal{P})$ denotes the non-empty ordered sequence of atoms that result from this process. $atom(\mathcal{P})$ is the first element of this sequence, and $alts(\mathcal{P})$ the remaining ones.

We distinguish two important cases:

- If $alts(\mathcal{P})$ is empty, we have a *strict* proof of $atom(\mathcal{P})$ on the basis of the preconditions. If particular, if $atom(\mathcal{P}) = \bot$, this proof denotes a conflict.
- Otherwise, we call the proof structure a *hypothesis*.

We define the set of hypotheses $H(\mathcal{P})$ that a proof $\mathcal{P}$ relies on as follows:

$$H(\mathcal{P}) := \begin{cases} \bigcup_{p' \in pre(\mathcal{P})} H(p') & \text{if } alts(\mathcal{P}) = \emptyset \\ \{atom(\mathcal{P})\} & \text{otherwise} \end{cases}$$

The function $H(\mathcal{P})$ associates with each proof structure the set of assumptions that the reasoning process has taken in its construction. This function is useful in the reasoning process, but also in the processing of the proof structures within the dialogue manager.

5.3.2 The Reasoner State

This section describes the state structure of the reasoning engine CIDRE. The reasoning state captures all the information about a reasoning process at one point in time. During this process, the reasoning state structure progresses from an initial state through the application of the iterative function (which is described in the following section) to a final state in which either the inconsistency of the input has been proven or all the models have been output.

The reasoner state can be separated into a static and a dynamic portion. The former contains data that is not modified by the reasoning process but is needed for processing. This portion mainly consists of a representation of the input theory.

The latter part essentially consists of three interrelated structures: the set of *usable facts* that have been proven already, the *agenda* of new items waiting to be combined with the usable facts, and the stack of *alternative states* that represents the topology of the complete tableau from the perspective of one node in that tableau. As such, the reasoning states correspond to the nodes in the tableau.

The reasoner state has the following components:

- *Usables*: Each reasoning state contains a set of proofs, called *usables*. This set represents the facts that can be used to generate new inferences by applying rules from the input theory. The proof structures contain information about how the fact was derived in the reasoning process. Each proof structure contains references to the proofs that were used as preconditions in the rule application. At the end of the reasoning process, the content of the usable facts determines the resulting model. The union of the hypotheses $H(\mathcal{P})$ of all usable proofs $\mathcal{P}$ is referred to as the reasoning state's set of hypotheses.

  The usable proofs are internally stored in the so-called *proof database*, a structure to maintain all inferences produced so far in this branch. The proof database is a monotonically growing set of proof structures and is reused if the current branch is closed. Thus, the proof database contains information on which assumption sets that have already been proved inconsistent (i.e. nogoods). The reasoner state also includes branch-specific access functions and indexes to the proof database.

- An *agenda* of new proof structures that are waiting to be processed: This structure is used to temporarily store proof structures that were created and that need to be combined with the usable facts. The agenda is a queue of proofs that have been derived on the basis of the currently usable facts. To emphasise that these proofs have not yet been processed, we also refer to them as *items*. The essential difference between the agenda and the usables is the fact that the agenda conceptually contains disjunctions which need to be disambiguated by branching before they can be added as atoms to the usables.

- A stack of *choice points* represents, informally, the rest of the tableau as seen from this reasoning state. It contains the possible backtrack places, i.e. alternative states to explore if the current state is found to be inconsistent. Essentially, this stack consists of information about which assumption was taken and which alternative assumptions could still be considered. Assumptions are taken when a non-unit disjunction is proven. In order to enable backtracking, for each of these assumptions copies of the reasoner state before the assumption was added is stored. The stack of choice points can be viewed to represent the tableau topology by linking the current reasoner state to alternative reasoner states (nodes).

The initial dynamic portion of the reasoner state consists of an empty tableau and a single item $\rightarrow \top$ on the agenda. The input theory has been preprocessed in a way that this single item will trigger all the facts from the input theory and subsequently all rules from the input theory will be applied to build new proofs.

5.3.3 The Iterative Reasoning Function

This section describes the core function for implementing the reasoning process. This function operates on the reasoning state structure described in the previous section. It computes a new state, which can be iterated until a fixed point is reached. Using an iterative function in this manner is the basis for enabling interactive use.

The basic reasoning algorithm works as follows: It processes the agenda of new items which have to be combined with all relevant usable proofs. In each application of the iterative function, one item is removed from the agenda, and combined according to the rules in the input theory with all existing facts (including itself). Subsequently, all new results of these combinations are added as new facts to the agenda, and the process is iterated. This corresponds to the PUHR rule. However, the situation is a little more complex: the results of combining facts may not be just simple facts, i.e. units, but disjunctions of facts (corresponding to the alternative consequences of a rule). Thus, the agenda has to store disjunctions as items. After removing a disjunction from the agenda, the first task is to determine if it is the special case of an empty disjunction. In this case, a conflict has been derived, and the assumptions leading to this empty disjunction constitute an inconsistent set. Otherwise, if the disjunction consists of only one item, proceed as outlined above. Otherwise again, the disjunction has two or more elements. Each of those, in principle, has to be processed in the same way. Each element constitutes an assumption or hypothesis that has to be evaluated. Therefore, the processing creates *branches* and proceeds in parallel for each of the alternative assumptions. This corresponds to the splitting rule in the PUHR method (cf. Section 3.6). Branches represent optimistic decisions where the reasoner decides to assume one alternative without a proof from existing knowledge. It may happen that the assumption turns out to be wrong (i.e. a conflict is derived, as outlined above), and one of the other alternatives is the right one.

Figure 5.1 illustrates the main iteration step implemented in the reasoning engine. This iteration is a function that computes the transition from the current to the next reasoner state. The reasoning procedure terminates when this function reaches a fixed point.

As illustrated before, the dynamic part of the reasoner state contains two main components, a set of usable proofs and the agenda. The latter contains new items that, one after the other, need to be combined the existing knowledge represented in the former, the set of usables. Here, all combinations of the new item and the existing usables have to be checked as to whether they can generate new inferences when used as preconditions in the rule set. Essentially, this means that first the set of rules that contains a precondition that unifies with the new item is determined. Then, for each element of this set the usables are used to unify with the remaining preconditions. If this process is successful the resulting postconditions of the rules are added as new items. The reasoner also contains a current set of hypotheses and a stack of choice points for backtracking.

The iterative function proceeds in the following phases (cf. Figure 5.1). If the reasoning state (S_0) is currently known to be inconsistent (i.e. a conflict has been proved based on the current set of assumptions) and it is possible to backtrack to a different assumption set, the backtracking is applied (Lines 1 and 2). It is possible to backtrack if and only if the current set of assumptions is non-empty. If so, there is a non-trivial disjunction that caused the one of the assumptions in the current assumption set. Backtracking eliminates this assumption and invokes the next possible alternative as present in the disjunction.

Function reasoningStep : $State \rightarrow State$
Require: $S_0 \in State$
  1: **if if** isInconsistent S_0 **and** canBacktrack S_0 **then**
  2:     backtrack S_0
  3: **else if** isInconsistent S_0 **then**
  4:     reportFailure S_0
  5: **else if** hasEmptyAgenda S_0 **then**
  6:     reportSuccess S_0
  7: **else**
  8:     $(item, S_1) =$ popItemFromAgenda S_0
  9:     **if** isAlreadyKnown $item$ S_0 **then**
 10:         s1
 11:     **else**
 12:         $(\mathcal{P}_{new}, S_2) =$ addToUsables $item$ S_1
 13:         $newItems =$ getCombinations S_2 $\mathcal{P}_{new}$
 14:         addToAgenda $newItems$ S_2
 15:     **end if**
 16: **end if**

Fig. 5.1. The main iterative function in the reasoning process.

If the state is inconsistent and the current set of assumptions is empty, the inconsistency of the input specification has been proved, and thus the reasoner can return by reporting failure (Line 4).

Otherwise, the current state S_0 is not known to be inconsistent. The next check concerns the agenda of items that still need to be processed. If this agenda is empty, there are no more items to consider, and thus the current state represents a consistent solution. In that case, success will be reported for the current set of assumptions (Line 6).

Otherwise, there is an *item* (a proved disjunction) that can be popped from the agenda of $\mathcal{S}_0$ (Line 8). Since the agenda is part of the reasoner state, a new state instance is derived (referred to by $\mathcal{S}_1$). This state is then used in the further processing. Now the item to be processed has to be inspected. If it is already known to the reasoner (i.e. an equivalent proof is already in the set of usable proofs), no new inferences can occur (all inferences based on this item already been performed before) and thus the function returns without further processing (Line 10).

Otherwise, the item is added to the set of usable proofs for the state $\mathcal{S}_1$ (Line 12). This creates a new state ($\mathcal{S}_2$) and a proof structure which is returned as $\mathcal{P}_{new}$. If it is the atom $\bot$, the current reasoning state has now become inconsistent. Otherwise, non-unit disjunctions have been handled by creating new choice points for backtracking in $\mathcal{S}_2$. In effect, the creation of a choice point means that it is possible to re-start with an alternative (obtained from the disjunction) if the currently chosen element should fail. In the next step (Line 13), the combinations of the new proof with the set of usable proofs are generated (*newItems*). The new combinations are (possibly non-unit) items and have to be added to the reasoners agenda (Line 14). This results in a new reasoning state, which is finally returned. The next sections present different aspects of this core iterative function in more detail.

5.3.4 Rules of Inference

This section discusses the inference rules used in the reasoning process. The rules of inference apply at different stages in the iteration presented in the previous section.

Forward inference. Forward inference is the main inference rule in the reasoning process. It is always applied when a new item from the agenda is combined with the existing usable proofs in order to derive new consequences. The body of a clause is matched against the usable facts. This matching is formally defined as the calculation of most general unifier (cf. Section 3.2) of the rule body and elements to be chosen from the set of usable facts. Range restriction guarantees that the resulting head of the rule is a disjunction of ground facts (i.e. containing no unbound variables). Clause compilation can help to make this matching more efficient (see Section 5.3.6). In particular, the fact that one of the usable proofs is "new" in this incremental process is taken into account. That is, only inferences involving this new item are tried as candidates in the matching process. As a result of the forward inference process the matched facts (respectively their proof structures) and the head of the clause are stored in an item proof structure. An item represents a disjunction of alternative hypotheses. These alternatives are disambiguated by branching when the item is popped off the agenda.

Backward inference. Backward inference is a special inference rule that is applied when reasoning states that were generated as alternatives due to a disjunction in a parent state have been detected to be inconsistent. In this case, and all its alternatives have been proved inconsistent as well, then it can be inferred that the parent reasoning state is also inconsistent. For instance: if in a reasoning state $a \lor b$ must hold, but either is found to be inconsistent, then the reasoning state itself is inconsistent. Furthermore, once a is found to be inconsistent in the example, b can be seen as a direct proof because $a \lor b$ has to hold.

More formally, consider a clause $p \to q_1, q_2, \ldots, q_n$: If there are $n - 1$ conflict proofs corresponding to the q_i alternatives, each proof with additional assumptions a_i, then the following holds:

$$p \land a_1 \land \ldots \land a_{n-1} \to q_n$$

The remaining q_n is inferred in the current state with the given preconditions (rather than being assumed). If this inference again leads to an inconsistency the current state is closed and the above procedure is applied recursively in the backtracking process.

Nogoods. *Nogoods* are sets of facts found to be inconsistent during the reasoning process. The idea is that these sets of inconsistent assumptions can prune those parts of the search space that extend the assumption set found to be inconsistent. It is worth noting that in the search process a reasoner state may be based on a relatively large set of assumptions whereas only small portion of these may lead to an inconsistency. In fact, if this were not the case, the recording of nogoods should be avoided because exactly the same set of assumptions is unlikely to be re-visited again. However, since the detection of an inconsistency results in a proof structure, we can use this to determine the exact (sub) set of assumptions that are inconsistent. The smaller this set, the more likely it is to be found during the further search.

To formalise the generation of nogoods, we note that: If a strict proof $\mathcal{P}$ results in the singleton $\bot$, a contradiction has been found. In this case the proof's set of hypotheses $H(\mathcal{P})$ is a nogood. Nogoods can be viewed as clauses of the form

$$A_1 \land \ldots \land A_n \to \bot$$

where each of the A_i is a ground atom. Once nogoods have been inferred, they are stored and are checked later to see if the same (or larger) set of assumptions is generated again in a new proof. In that case, the reasoning engine can use the nogood to infer a contradiction.

In principle, nogoods may help to significantly reduce the amount of search by pruning whole branches of the search space. However, their management also requires some effort that increases substantially as the number of nogoods retained and it can not be guaranteed that the nogoods will actually be used. Therefore, the use of nogoods is an optional feature of reasoning engine. For instance, in the TRAINS domain the main search strategy is to increase an overall time limit, starting at one hour and relaxing it by one hour if no solution can be found. In these cases more analysis is necessary to determine how many nogoods are actually reusable. In particular, the only nogoods reusable across time limits are the ones that have been inferred independently of the current time limit.

5.3.5 Handling Equality

This section discusses the handling of equalities in the reasoning engine CIDRE. The handling of equalities significantly influences the behaviour and the implementation of the reasoning procedure. The PUHR method discussed in Section 3.6 assumes a Herbrand universe. This implies that terms used in the reasoning process are interpreted as themselves in resulting models. For instance, it is impossible for term $f(x)$ to refer to the same domain entity as term y. Concerning the modelling, this constitutes a restriction, in particular, since Skolem function terms are used in place of existentially quantified variables.

An approach to avoid problem arising from this restriction is the Extended Positive Tableaux approach discussed in Section 3.6. Here, when introducing a new term due to an existential quantification, all possibilities to use existing terms are considered before introducing a new term, which is then known to be distinct from all others. This allows to make the Herbrand universe assumption. The advantage relies in the fact that there is no need to deal with equalities at a later stage, thus the processing may be more efficient in that respect. Also, generation of minimal finite models can be guaranteed. The disadvantage lies in an unnecessarily early commitment to equality assumptions.

We aim to avoid these restrictions in our reasoning engine. Our approach attempts to handle equalities in a different way, which can be seen as a compromise between the two approaches discussed. We allow terms to become equal at a later stage in the reasoning process avoiding a premature commitment to equality assumptions. However, this also incurs some complications: if $f(x)$ can become equal to y, also other terms like $g(f(x))$ and respective atoms have to be normalised, potentially resulting in new items that need to be considered. Function terms may be viewed as placeholders that can be proved identical to some other term in the reasoning process. To ensure that functions behave as expected, certain preprocessing tasks need to be performed on the input theory: For each n-place function term $f(t_1, \ldots, t_n)$ the reasoner automatically generates a $n+1$-place predicate f, constrained by the following axiom:

$$f(t_1, \ldots, t_n, x) \land f(t_1, \ldots, t_n, y) \rightarrow x = y$$

The idea is to enforce equality (the behaviour of f as a function) by the regular inference mechanism. To this end, whenever a new term $f(t_1, \ldots, t_n)$ is introduced by a proof $\mathcal{P}$ (first appears during the reasoning), a fact $f(t_1, \ldots, t_n, f(t_1, \ldots, t_n))$ is automatically generated by the reasoner (with the same preconditions $pre(\mathcal{P})$).

In any reasoning state, a defined set of equalities is known. These may be used to define the *normal form* of a proof with respect to this reasoning state, i.e. the proof with all possible term simplifications applied. Equalities are interpreted as mappings from more complex terms to simpler terms, as defined by the term order $<_T$. We introduce a function $NORM$ that normalises a given term according to a set of usable facts (proofs) and returns the normalised term and the proofs used in the normalisation. Recording the proofs is necessary in order to make the normalisation a valid proof step (i.e. to ensure that all preconditions are recalled).

The function $NORM$ performs the normalisation of a term $t = f(t_{1..n})$ which is composed of a functor f and a (possibly empty) list of subterms $t_{1..n}$ in the following steps: If there is a proof for an atom $f(t_{1..n}, x)$, where x is some term, then x is the normalisation of t and this proof is returned. Otherwise, the terms $t_{1..n}$ are normalised recursively, noting the proofs used. If the resulting term t' is different from t, proof search step (for $f(t'_{1..n}, x)$) is re-performed on it. The respective result and all proofs are returned. The normalisation of terms can be used to define the normalisation of atoms in a straightforward way.

In order to be able to also use terms that behave like terms in a Herbrand interpretation, we introduce a special sort of terms, called *extensional terms*: these are integer numbers and strings written in double quotes notation, for instance, "Avon". Extensional terms are interpreted as themselves and cannot be normalised. This information can be used efficiently in the reasoning process, for instance, when an equality of the form $1 = 2$ has been derived, it can be instantly reduced to $\bot$.

5.3.6 Optimisation Techniques

During the reasoner process a considerable effort is spent for finding new combinations of usable facts to fill the preconditions of rules. Thus, it is important to maintain the set of usable facts in data structures that make related queries (and updates) as efficient as possible. In the reasoner two related strategies are implemented: At run time, usable facts are indexed by their argument terms in different argument positions. This indexing enables efficient access if some arguments are already known and some are to be matched This is the common case in the process of matching the preconditions: in each match, some variables will be instantiated and those that are already filled are used to constrain the search space.

In addition, during the initialisation of the reasoning process, the domain theory is preprocessed in the sense that the rule bodies are compiled into a specialised form that takes into account the order in which variables are filled. In that preprocessing, specialised search instructions based on which variables are known to be filled in which match are generated. This is described in more detail in the following.

In principle, in the reasoning process matching a rule body (i.e. a set of precondition atoms, each of which containing a set of variables) is the recursive process of determining a most general unifier: Initially, all predicate arguments will be variables. In each subsequent step variables will be instantiated, such that, at the end of the matching process, a list of variable-to-term assignments is generated, where each resulting atom is in the usable facts set. Consequently, in each match step (as long as there are ones waiting to be matched), first, one precondition atom is chosen. Then, for this predicate the list of all candidate usable facts is determined. The candidates have to match the current variable assignments. Finally, for each candidate in turn, The variable assignment is updated and the match process is repeated with the remaining preconditions.

The key to optimising this process is that the number of alternative instantiations (candidates) should be minimised. We aim at achieving this by trying to find the atom pattern with the smallest number of unbound variables. Atoms with fewer unbound variables will result in fewer candidate facts from the set of usable facts. Once no unbound variables remain, the atom is ground. In this case, the process will determine at most one candidate by detecting whether the respective atom is contained in the set of usables.

The optimisation approach is to statically compile rule bodies into an intermediate representation which allows an efficient processing at run time. A clause body is represented as a list of atoms. We can assume that the terms in the atom argument list are only variables. Otherwise, a preprocessing is performed in the following way: If constants are among the terms, a new auxiliary variable and a special atom is added to the body which will instantiate the new variable. Similarly, if function terms appearing in the atoms, they are replaced with additional atoms. This is illustrated in the following example: $p(f(x), 1)$ is translated into

$$p(v_0, v_1) \wedge f(x, v_0) \wedge const_1(v_1)$$

Here, $const_1/1$ is the special predicate that only holds for the constant 1.

The target representation is a list of Match operations, such that each works as follows: a `Match(pattern)` operation is a primitive match of an atom against the set of usable facts at run time. Here, the pattern is a list of flags, which is intended to mean the following: If the flag is

- `IN`, the respective variable has already been bound to a value, and this value can be used when accessing the database of usable facts,
- `OUT`, the respective variable has not yet been bound, and will be bound by the candidate usable facts,
- `Check` i, the respective variable has not yet been bound, but appears more than once in the current atom (e.g. $p(x, x)$). In order to guarantee that only those candidates that return the same binding for x are used, the pattern for $p(x, x)$ with x unbound, will be `[OUT,CHECK 0]`.

The algorithm in Figure 5.2 translates a rule body into a sequence of matching operations. The base case of the function mOps just returns an empty sequence of operations if there are no more atoms to match. Otherwise, the atom with the least number of OUT variables (given the current set of already instantiated variables, which is initially empty) will be determined and an individual Match operation will be constructed. The remaining operations are determined recursively for the remaining atoms to be matched.

Function mOps : $[Variable] \times [Atom] \rightarrow [Match]$
Require: $v \in [Variable]$, $a \in [Atom]$
1: **if** isEmpty a **then**
2:     return []
3: **else**
4:     let $(a', v') = $ the atom with the least number of unbound variables
5:     let $p = $ the pattern instantiating v' in a
6:     return Match$\{a,p\}$: mOps$(v \cup v', a\backslash\{a'\})$
7: **end if**

Fig. 5.2. Translating a rule body into a sequence of match operations.

In addition, given a rule body consisting of atoms $a_{1..n}$ in the input theory, for each atom a_i one specialised variant of the body will be generated using the mOps algorithm where a_i is required to be the atom instantiated first. This is done because in the reasoning algorithm the rules are applied in situations when one new atom has to be combined with the existing set of usable atoms, i.e. this (ground) atom will always be fully instantiated. A further optimisation consists in using knowledge about functional predicates, i.e. predicates that will return at most one binding in their last argument position.

A complementary optimisation strategy concerns the agenda, i.e. the run time queue of items to be integrated in the reasoning process. These proofs have been inferred on the basis of the currently usable facts and assumptions. When a non-unit (disjunction) is to be integrated, it is possible to search each of the elements as strict proofs in the agenda. If found, the disjunction is not necessary since it is disambiguated by existing inferences. However, there is a trade-off concerning the effort spent for searching.

5.4 The Interactive Protocol

So far, the reasoning has been described as an autonomous process that reads an input specification and produces a stream of inferences by repeatedly applying the iterative function introduced in Section 5.3.3. However, our goal is not to build an autonomous reasoning engine, but an interactive one. To this end, we have designed an *Interactive Protocol* and implemented it as an integral part of our reasoning engine CIDRE. The main tasks of the interactive protocol are to enable an external component

- to access to database of generated proof structures, and nogoods.

- to manage a set of concurrent reasoning processes (incrementally create tasks, i.e. create specialisations of existing tasks which take advantage of the inferences already produced in the base task).
- to query the current state of reasoning process (proofs, branches, conflicts).
- to apply domain-specific heuristics for guiding the reasoning process (agenda order, branch selection, search strategy).

The interactive protocol contains both access methods (or request message types) and an event listening mechanism (or event messages). The event-based mechanism allows a *client component* to create a trigger to react to a specific situation during the reasoning process. For instance, the trigger may become active when a branch is created in the reasoning process.

The reasoning process maintains an internal log. It is used, for instance, to note branches and some kinds of events that need to be documented. In long running processes, this may consume a considerable amount of memory. Thus, a part of the interactive protocol is defined in a way that filters the log for entries it returns and removes them subsequently. Alternatively, the log can be emptied explicitly. Testing the interactive protocol is described in Section 5.5.6.

5.4.1 Accessing the Proof Database

Access to the proof database means that an external client can perform queries concerning the proofs that have been generated. In particular, proofs can be inspected in a detailed way. For instance, the proof structure can be browsed for antecedent proofs. Furthermore, the database can be queried in order to determine if a certain set of ground atoms has been proved inconsistent (is a superset of a nogood). Finally, proof database mechanisms can be used to perform term normalisation in order to apply derived simplifications to a given term. This is useful for determining if two different terms are mapped to the same entity.

5.4.2 Managing Concurrent Reasoning Processes

Managing concurrent reasoning means that an external client can initiate and manipulate reasoning processes which execute in parallel to potentially existing ones. In order to do so the following message types have been devised:

CREATE_TASK: this message creates a task with set of assumptions, optionally based on an existing task. This message creates a new reasoning task based on an existing task with additional assumed facts. The reasoning in new task continues where the existing task stopped and is re-activated by adding the new facts to the agenda. In particular if the base task stopped in a model found with an assumption stack, the new facts may contradict these assumptions and therefore backtracking must be possible. The newly created task starts immediately in a parallel processing thread.

SUSPEND_TASK, RESUME_TASK: These messages can be used to suspend and resume the respective reasoning task.

DISPOSE_TASK: This message stops the respective task and frees any reasoning engine resources associated with it.

QUERY_TASK_MESSAGES: This message can be used to obtain messages from the reasoning engine. A filter pattern can be provided to restrict the transmission to certain types of messages.

WAIT_FOR_TASK: This message is used to wait for a task to finish.

Concerning incrementality, one has to distinguish three factors: whether the reasoning process has finished, the number of models that have been found, and what kind of modification is to be applied to the input. One of the most common situations is that a ground fact will be added to a finished reasoning process. For instance, in a dialogue situation this is the case when the system has found a solution and the user asserts a new requirement.

In principle, incremental changes should come up with the same result that a non-incremental reasoning process (i.e. one restarted from scratch) would come up with. Different situations can occur on incremental changes: When a new fact is introduced, it can be added to the reasoner's current agenda. However, the protocol must make sure that this fact is still there when the reasoner decides to backtrack, i.e. the stack of choice points has to be adapted. The same applies to the addition of a non-unit disjunction.

Adding a negated ground literal is even easier, since only the presence of the respective positive unit has to be checked. Adding a complete clause constitutes a more substantial change to the reasoning process. In this case, the protocol implementation needs to combine all existing facts and assumptions with the new clause.

Similarly, when an atom or clause is removed, the reasoning engine can, in principle, detect in which situations (if any) the clause was used so far. The respective inferences and branches can be removed. However, this may lead to re-activating states that became inconsistent because of this atom. In practice, the reasoning process is re-started in this situation.

5.4.3 Controlling the Reasoning Process

Controlling the CIDRE reasoning process is possible through installing event listeners. This is useful for obtaining specific information without polling and for specific extensions of the reasoning strategy or heuristics that may guide the reasoning process. One such heuristic for long running reasoning processes may, for instance, prefer the processing of proofs related to certain terms that are known to be in focus in the current dialogue situation. It is also useful for implementing access to external knowledge sources, which is discussed in the Section 5.4.4.

The event messages defined for controlling a reasoning process are:

FINISHED: This event signals that the reasoning process has finished.

NEW_SOLUTION: This event signals that the reasoning process has found a new model.

NEW_CONFLICT: This event signals that the reasoning process has detected an inconsistency and that backtracking will occur, if possible.

NEW_TERM: This event signals that a formerly unused term has appeared in the reasoning process.

PUSH: This event signals that an item is going to be pushed onto the agenda. The client can influence the place (order) that the item is going to be assigned on the agenda.

POP: This event signals that an item is going to be popped from the agenda.

NEW_BRANCH: This event signals that a new branch is to be created because a disjunction was popped from the agenda.

5.4.4 Accessing External Data Sources

Access to external data sources can be implemented on the basis of the interactive protocol. This section discusses various kinds of external data sources and how they can be integrated with the reasoner.

One of the most promising options is to implement *external procedures*, i.e. procedures available in a native implementation that provide certain domain-specific inferences. For instance, a native navigation service may provide inferences concerning the travel durations of routes once these are described to the extent required by the service. This usually means that the information concerning the source and destination needs to be sufficient to map these to the internal representation of the services. However, the service might also be able to provide inferences even if only part of the information is available, for instance, the country or province. In that case, the inference provided by the navigation service should be a lower bound that is refined when more information becomes available. Such inferences must be proof steps, i.e. the arguments the result depends on need to be made explicit. Thus, an inference resulting from an external procedure, like inferences produced in the reasoner core, represents the antecedent proofs it is based on. Possible other function candidates include regular expressions and format conversions for temporal expressions.

Textual file contents can be represented as numbered lines of a file. Alternatively, the file content could be represented as a list of strings (not implemented). Concerning the special case of XML-structured file contents, implementing a theory for XML concepts, such as elements, attributes, nodes etc. seems an adequate approach. Calendar files are, in principle, another specialised file format. However, in this case a calendar access application programming interfaces (API) may be used.

Access to relational databases is an interesting option for external data access. It may be approached in different ways. For instance, the reasoning process could use a relational database for storing static relational data (read only access, access could be performed efficiently using specialised SQL queries). Writing to a relational database has not been considered because a database can only represent a set of facts (i.e. not a tableau of alternatives), and it can only represent extensional terms. A different approach is to *emulate* a relational database (cf. Section 4.4.7), i.e. to translate the database schema and the content to a domain theory. In that approach, access to the real database is not necessary in the reasoning process. However, one may take advantage of the existing static data and modelling in the relational database.

5.4.5 Adapting the Search Strategy

The default behaviour in our reasoning engine CIDRE is a depth-first search. However, in the following we discuss how this behaviour can be influenced using the interactive protocol.

As outlined in Section 5.3.3 the reasoning process performs a search in the space of hypotheses sets, adding hypotheses when disjunctive items are branched and replacing them when a hypotheses set leads to an inconsistency. Different search strategies can be implemented in the reasoner. These concern the way (a) how to switch between alternative reasoning states (i.e. reasoning states that were created by assuming alternative elements of a disjunction), and (b) how backtracking is performed after a set of assumptions has turned out to be inconsistent. Viewed from the tableau perspective the situation is as follows: The tableau being built in the reasoning process is a tree structure whose leaves are expanded according to the tableau construction rules, until a saturated leaf is found or the branch is found to be inconsistent. The question of deciding in each step which leaf to expand can be viewed as a search process.

The first option is to use depth-first search. When an existing reasoning state (tableau node) is branched into new states, one of these is visited before all of the other alternative states. One variation of this search strategy is bounded depth-first search. Here, depth-first search is performed up to a maximal depth. If this depth is reached, the alternative states are visited. The depth is increased once no state of lower depth is available. The alternative approach is breadth-first search. Here, the new states created from an existing one are visited after all of the other pre-existing states have been visited. A practical problem arises from breadth-first search, namely, to maintain a large number of states in memory at the same time.

Comparing the strategies, we note that depth-first search typically allows more concise algorithms and can be less memory consuming. Breadth-first search, on the other hand, is sometimes necessary to avoid following infinite branches (dead ends), or to guarantee that the shortest path is found. Bounded depth-first search aims to achieve depth-first efficiency, while avoiding infinite dead ends by a bound that can only be increased if the complete search space below this bound has been explored. In our case, a breadth-first strategy would expand the tableau tree structure level by level, whereas the depth-first strategy always expands the same branch.

5.4.6 Implementation

Our implementation of the interactive protocol has the following characteristics: Each reasoning process is a separate thread of execution in the reasoning engine. The communication takes place through concurrently accessible variables and queues. The behaviour of the data structures can be described in a state transition network. At any time, a concurrent variable is either EMPTY (not containing a value) or FULL (otherwise). In contrast to a variable, a concurrent queue may store a sequence of values. The operations are thread-safe, so for write accesses, no losses are possible. Writing a concurrent variable blocks the writer if it is not empty. In the case of a queue, the writer is not blocked and the value is stored. Read accesses can be blocking or based on a specified time-out. The data structures can contain arbitrary message objects, since the processes run in the same address space.

The communication uses the following instances of the concurrent data structures:

- Reasoner-to-Client queue: this queue contains the messages produced by the reasoner.
- Client-to-Reasoner queue: this queue contains the control messages sent to the reasoner. Most importantly, this is used for stopping or suspending the reasoner.
- For external communication, i.e. with a client of the reasoner (e.g. dialogue manager), two queues are used for communicating text-based messages.

For communication with external processes (and, for instance, other programming languages) a text-based message bridge is provided. One important aspect of using a text-based communication format is the requirement to flatten the complex data structures (e.g. proof structures) used in the reasoning process. In particular, references to shared data structures (e.g. inferences communicated at one point in time and later referenced in some other inference) need to be expressed via ID values. This introduces the requirement for registering proofs in order to detect which proofs have already been used etc. Fortunately, the proof database performs the registration of proofs and the assignment of ID values. Also, since the reasoning process is monotonic, the structures once identified with an ID value are guaranteed to remain unchanged.

5.5 Development and Testing

This section mainly presents experiences made using the reasoning engine CIDRE as an autonomous system. Although CIDRE was not designed as a completely autonomous process, a certain level of autonomy is desirable. The system should be able to derive solutions for relatively simple problems without requiring interaction. Also, the reasoning engine was used in this way to support the development of the domain models.

It was however not the primary goal to build a reasoner is fast in finding models, but one which provides the kind of information we require in our proposed dialogue management approach. In fact, it is not only required to know if a model exists our what its content is, but also how it has been generated.

In order to evaluate CIDRE as an autonomous reasoner, tests from the literature have been applied and corpus of test problems based on our library of domain models has been designed. The latter is arguably more realistic for applications. The tests are an important requirement for regression testing, i.e. the developer needs to be able to verify that new versions of the reasoner or the domain theory library produce the expected results.

In the autonomous mode, the reasoner reads its input from a test specification file containing both the axioms from the domain theory used and the specific test case. It generates a log file that contains the relevant information concerning the reasoning process, most importantly, if the model generation was successful or not. The format of the log file generated by the reasoning process can be configured with a number of options. The principal contents of the log file are:

- *agenda selection*: This option documents which items are chosen from the reasoner's agenda (its current heap of items to be processed).
- *inferences*: The inferred proof is documented. Usually, the head of the proof is noted together with the rule ID and ID values referring to the proofs used to fill the rule's preconditions.
- *branches*: Importantly, branches are documented in the log format. A branch occurs if a proof with a non-trivial disjunction is chosen from the agenda. In this case, a new assumption will be introduced to the subsequent reasoning.
- *conflicts*: Conflicts are situations in some sense similar to branches, since they change the current assumption set of the reasoner state. Furthermore, the rejection of an assumption influences the behaviour of its alternatives in the assumption's originating disjunction.
- *solutions*: Finally, if solutions are found their details are listed for efficient comparison.
- *timer values* are included in the log trace to be able estimate the performance of the process.

One of the major difficulties working with the reasoner in autonomous mode was the sheer size of the log files produced by some test cases. Since the proof and branch structures produced are inherently complex, the log file is rather verbose. Strategies applied to tackling these issues included directly writing compressed log files and producing different log files based on different filter criteria in parallel. In addition to that, a set of off-line tools has been developed for reconstructing parts of the reasoner state or complete proof traces from log file contents.

5.5.1 Performance Tests

Bry and Yahya have defined four benchmark suites, referred to as A, B, D and F, for evaluating MM-SATCHMO, a model generator based on the PUHR method (cf. Section 3.6). The benchmarks consist of up to 100 000 clauses and give rise to as many as 100 000 minimal models.

The suites are defined as follows (in a slightly adapted notation for consistency with our notations):

$$A(n,m) := \{\top \to a[i,1] \vee \ldots \vee a[i,m] \mid i = 1, \ldots, n\}$$
$$B(n,m) := C(n,m,m^n) \cup A(n,m)$$
$$C(n,m,k) := \{M[j] \to \bot \mid j = 1, \ldots, k-1\}$$
$$D(n,m,k) := E(n,m,k) \cup A(n,m)$$
$$E(n,m,k) := \{a[i,j] \to a[i+1,j] \mid i = 1, \ldots, n-1, j = 1, \ldots, k\}$$
$$F(n,m,k) := C(n,m,m^n) \cup E(n,m,k) \cup A(n,m)$$

In these equations, $A(n,m)$, $B(n,m)$, $D(n,m,k)$, and $F(n,m,k)$ denote the contents of clausal theories of the suites A, B, D, and F, respectively. C and E are used to construct auxiliary sets of clauses. $a[i,j]$ is used here to denote distinct atoms of arity 0 constructed out of the parameters i and j. $M[j]$ denotes the conjunction of atoms that make up the j-th model of suite A. Thus, the problems are essentially propositional. Suite A is defined to construct m^n distinct models, each consisting of n atoms. Suite B adds constraints that filters out all but one of these models. Suite D adds "chains" of implications of length k between the facts $a[i,j]$. Thus, the larger k, the more restricted the problem becomes. Finally, suite F is constructed by adding the filter clauses of C to D.

Times are reported for the following configuration: ECLiPSe Prolog 3.5.1, HP-UX 10.20 Workstation HP Visualise C 160, PA-8000 processor at 160 MHz, 192 MB RAM. The times obtained by [79] are summarised in Table 5.1[1].

We have performed the same experiments with our reasoning engine CIDRE. Our configuration consisted of an Intel(R) Core2(TM) Duo (2660 MHz, 2 GB RAM) running Linux 2.6.27. The times obtained in our experiments are summarised in Table 5.2. The times are user times that have been obtained by using Haskell's `System.Posix.Process. getProcessTimes` function. The system times are typically less than one percent, so the values are close to the actual clock time differences.

---

[1] For space reasons, we restrict our discussion to the case $n = 5$, which is the maximal value in [79].

m	3	4	5	6	7	8	9	10
$A(5,m)$	0.15	1.44	13.74	85.05	413.62	1 445.91	4 782.78	13 317.30
$B(5,m)$	0.10	1.29	14.99	102.43	451.75	1 799.27	5 737.51	15 398.20
$D(5,m,1)$	0.07	0.63	5.19	42.13	299.14	928.12	3 165.38	8 819.52
$D(5,m,\lfloor\frac{m}{2}\rfloor)$	0.07	0.29	1.84	4.38	28.98	57.45	315.45	542.96
$D(5,m,m-1)$	0.05	0.28	0.63	1.59	3.51	7.09	13.14	22.88
$F(5,m,1)$	0.12	1.40	16.60	100.38	462.95	1 688.92	5 704.90	15 242.60
$F(5,m,\lfloor\frac{m}{2}\rfloor)$	0.12	0.97	12.69	60.84	303.68	989.93	3 585.84	8 608.44
$F(5,m,m-1)$	0.08	0.59	6.06	32.30	134.27	482.85	1 498.90	4 202.49

Table 5.1. Summary of CPU times in seconds for MM-SATCHMO [79].

m	3	4	5	6	7	8	9	10
$A(5,m)$	0.04	0.18	0.64	1.49	3.34	6.73	12.87	22.46
$B(5,m)$	0.18	1.12	7.22	31.92	120.25	376.14	1 064.83	2 687.09
$D(5,m,1)$	0.01	0.08	0.27	0.80	1.99	4.29	8.30	15.24
$D(5,m,\lfloor\frac{m}{2}\rfloor)$	0.01	0.02	0.12	0.14	0.46	0.54	1.47	1.65
$D(5,m,m-1)$	0.00	0.00	0.01	0.01	0.02	0.02	0.02	0.02
$F(5,m,1)$	0.09	0.65	3.81	18.27	71.32	234.33	692.21	1 816.32
$F(5,m,\lfloor\frac{m}{2}\rfloor)$	0.07	0.41	1.36	3.79	8.50	17.42	33.71	61.83
$F(5,m,m-1)$	0.09	0.47	2.32	6.00	22.79	45.04	148.18	256.96

Table 5.2. Summary of CPU times in seconds for CIDRE.

The numbers are somewhat difficult to interpret. Firstly, both the hardware and software configurations are quite different. For instance, MM-SATCHMO is implemented in Prolog, whereas CIDRE is written as Haskell programme. In addition, the numbers are not directly comparable because MM-SATCHMO performs a considerable effort to report only minimal models. This is not done by CIDRE and may explain some of the time differences in A suite. However, CIDRE produces the same models for the suites discussed here. Another problem concerns the construction of the C set of clauses. In [79] the authors note that these constraints are expressed as Prolog clauses using Prolog's internal :- operator. Thus, these are treated differently from regular implication clauses. The C set used in B and F constructs as much as $5^{10} - 1 = 99999$ of these constraints. It seems that the use of the :- operator substantially improves performance, whereas CIDRE has to use its regular matching process to derive the conflicts. From our point of view, the C set uses a very special feature of representation, which is rarely seen in real life applications in this form. In addition, in contrast to MM-SATCHMO, CIDRE has to do some extra work in order to generate the proof structures to be used as part of the dialogue management approach. This may also incur some performance penalty in the general processing. In summary, we argue that the numbers give a rough indication that our implementation is efficient for regular tasks. In order to support this claim, we discuss some test cases constructed for testing our library of domain models and the TRAINS domain in the next section.

5.5.2 Test Cases

This section describes examples of the test cases that have been used both during the development of the CIDRE reasoning engine and the application domains.

Basic tests. For the reasoner development, a set of basic test cases has been designed to be used for regression testing. This mainly concerns the following fundamental functionalities:

- Disjunctions: for inspecting basic branching behaviour and selection of disjunction elements. They are also necessary for inspecting the handling of closed branches.
- Equality: for inspecting how equality constraints are handled, e.g. ensuring the transitivity of the equality relation. Also, the behaviour of the equality processing with respect to deeply nested terms needs to be checked.
- Nogoods: These tests concern the handling of nogoods. In particular, independent nogoods (i.e. nogoods that do not have common assumptions) need to be verified.

Library tests. Apart from the basic tests, more complex tests have been implemented in order to test the usefulness of the constructions provided by the library of domain models. Among the library tests, lists are one of the most important test cases, since they are relatively complex structures and are used in other models as a component type. In particular, the length of the list can be parameterised in a test case in order to analyse how the reasoning engine deals with larger structures.

We define a test suite for lists in the following way:

$$L(n) := \{list(l), \ length(l) = n\}$$

The CPU processing times for finding the first (and only) model are summarised in Table 5.3. The times have been obtained using the same hardware configuration as in Section 5.5.1. As can be seen from the numbers, espe-

n	0	1	2	3	4	5	10	15	20
CPU time [s]	1.61	1.68	1.70	1.70	1.75	1.78	2.07	2.60	3.37

Table 5.3. CPU processing times for $L(n)$.

cially column $n = 0$, there is some initial overhead that is required even if the problem is trivial. If fact this overhead would exist even if no actual problem was specified at all. The reason is that the system contains specifications in the arithmetics module which need to be evaluated. For instance, an initial table of entries modelling the $leq/2$ relation for the numbers provided in the arithmetics module will be computed. For the L suite, the numbers indicate that the test cases can be solved quickly. In fact, branching is not necessary here because the system can use the $sum/2$ relation from the arithmetics module to reduce the problem step by step in a deterministic manner. Nevertheless, some non-linear effort is necessary, for instance, for computing the $member/2$ relation. In summary, the data obtained point to the practicality of the modelling when the involved effort is kept in mind.

Another important reusable library is the fluent change events module. This library can be tested by constructing test cases that vary in the number of required fluent change events. We define a test suite for lists in the following way:

$$FC(n) := \{value_f(i, i) \mid i = 0, \dots, n\}$$

Since all the atoms $value_f(i, i)$ are pairwise distinct, n fluent change events are inferred.

The CPU processing times obtained are summarised in Table 5.4. In the

n	0	1	2	3	4	5	10	15	20
CPU time [s]	1.23	1.54	2.00	2.47	3.26	4.12	17.21	71.35	249.78

Table 5.4. CPU processing times for $FC(n)$.

FC suite, the situation is more complex than just inferring list structures. The reasoner infers $O(n^2)$ fluent changes. On that basis, it has to branch for hypotheses as to whether these changes are primitive or not. In the cases presented the reasoner is able to successfully derive solutions by assuming primitive changes and by showing that these lead to a saturated model. Although the numbers indicate a substantial complexity of our approach based on fluent change events with rising values of n, we interpret them to show that for a practical range of numbers, for instance, as occurring in the TRAINS domain, the approach is effective. To support this argument we discuss this domain as a test bed in the following.

TRAINS domain. The TRAINS domain has been one of the key domains of the reasoning engine development. For this reason, the largest number of test cases has been constructed in this domain. This part of the test library is used to test aspects of our implementation of the TRAINS world, i.e. the TRAINS domain theory described in Section 4.5. Two experiments based on this model will be discussed in the following sections.

For operational purposes the TRAINS domain model has been extended with the following axiom:

$$\top \rightarrow general\_time\_limit(1) \vee \ldots \vee general\_time\_limit(24) \qquad (5.1)$$

$general\_time\_limit(n)$ is further specified to restrict the occurrences of fluent change events to a time range of at most n hours of domain time. This essentially implements an iterative deepening scheme within the domain model, since the disjunction in 5.1 is the first to lead to a tableau branch if there are no others with a rule body of $\top$. This is guaranteed unless there are such disjunctions in the test case. One potential advantage of this scheme is that nogoods collected can be reused for multiple $general\_time\_limit$ values.

5.5.3 Enumerating TRAINS Models

In this section we describe how practically all the TRAINS models can be enumerated using our reasoner CIDRE. This is useful for obtaining an idea about the complexity of the domain and the performance of the reasoner when applied to the domain model as introduced in Chapter 4. By "practically all" we mean that there is no arbitrary upper bound to the number of models. However, for time and space considerations we limit our discussion here to the first 300 models, after which we stopped the model generation process.

The models were generated by adding the following problem specification

$$\top \rightarrow at(E_2, t, c)$$

Here, t and c are uninstantiated constants that create an ambiguity that the reasoner attempts to resolve by instantiating them to extensional values such as $t = 0$ or $c =$ "Avon". This axiom will only trigger models that involve engine E_2. However, in order to get all models, one may add respective axioms for E_1 and E_3. We have avoided this for the sake of the presentation.

The processing times obtained and other measurements are summarised in Table 5.5. The CPU times are from one instance of a reasoning process and represent the times required to find the n-th model (including any previous models). The size numbers report the size of the resulting model in terms of ground atoms. The d-hours column reports the value of $general\_time\_limit/1$ used. Based these values, we have selected the first and last models in the table as long as they were within our limit of 300 models. The last model of the experiment has also been included. The depth value range indicates the number of internal assumptions that were used to find models with the specified range of models. Finally, n_{inc} reports the number of inconsistent models produced before finding the n-th model.

From Table 5.5 we can draw the following conclusions: The process takes a certain time before the first model is found. This is time is mostly required to process the problem-independent information and might be reduced by an optimisation in the sense that these inferences could be pre-computed. With

n	1	2	7	8	62	63	300
CPU time [s]	52.85	56.31	165.01	167.89	4 242.39	4 248.99	287 047.11
Size	7 567	7 567	7 682	7 567	7 823	7 567	7 963
d-hours	1	2	2	3	3	4	4
Depth	1	1–9			1–22		1–34
n_{inc}	0	6	22	35	996	998	8 466

Table 5.5. Measurements for enumerating TRAINS models.

a rising value of *general_time_limit* the number of models rises extremely quickly. This is mostly due to the vast number of possibilities of attaching combinations of boxcars in each transportation event. The number of options is not infinite, though, since the boxcars must be present at reachable locations. If only aspects of the solutions are considered, such as where the transported goods end up eventually in the scenario, then a substantial number of these models will be equivalent. Another form of redundancy is possible in the current modelling in the sense that models requiring less domain time are repeated when the domain time constraint is relaxed (cf. columns $n = 1, 2, 8, 63$). This may be addressed by a tighter formulation of the axioms concerning *general_time_limit*/1. Furthermore, one can see from the data that while the first models can be found relatively quickly, processing time increases substantially in the further processing. However, the size of the models increases only moderately.

In summary, we can note that it is, in principle, possible to enumerate all the TRAINS models. If an upper bound is imposed on the domain time allowed, then the number of models will be finite. However, this is a rather theoretical observation, since the number of models will be beyond practical use. Instead, we argue that additional constraints need to be added to reduce the search space. Additional problem-specific constraints can only improve on this baseline in the following sense: Filtering the enumerated models according to additional constraints is trivial. Thus, additional constraints will not increase the search. On the other hand, of course, the time required to find the first model that satisfies the additional constraints may take substantially longer. Nevertheless, we assume that additional constraints will be used more efficiently in the search than a mere filter. This is discussed in the next section.

5.5.4 A Sample TRAINS Test Case

The previous section described an approach to enumerate TRAINS models. In this section we describe a particular test case in more detail in order to illustrate that the reasoner is also capable of solving more complex cases than those that were discussed in the enumeration approach.

The sample problem presented here is to move one tanker of orange juice to Corning. This is encoded in the following specification:

$$oranges(or).$$

$$at(or, t, \text{``Corning''}).$$

$$value(j(or), t, 1).$$

Herein, the $j(or)$ denotes a fluent which indicates the state of the product, i.e. requiring that juice has been produced from it. The resulting model contains 8843 atoms, and its duration in domain hours is $t = 8$. The solution found can be summarised as follows (cf. Figure 4.4 for an illustration of the domain): In order to get orange juice to Corning, first the juice has to be produced in the factory at Elmira. To this end, oranges available at the warehouse in Corning have to be moved there. This involves boxcars (either from Dansville or Bath) and which to be pulled by an engine. The tanker in Corning also needs to be moved to Elmira in order to be filled with the juice. Thus, the chronological order of the events is: Move engine E_3 from Elmira via Corning to Dansville, attach boxcar B_6 there to E_3, take both to Corning, load B_6 with oranges, attach the tanker, take all cars to Elmira, produce the orange juice (moving the oranges from B_6 into tanker), attach tanker to E_2, take E_2 with orange juice in tanker to Corning.

Inspecting the solution found by the reasoner, it seems unclear why two different engine entities are involved in the solution while one would be sufficient. However, as attaching transport cars to engines does not incur any costs, the solution found is not worse to the reasoner than the solution involving only one engine. In fact, only the total time is modelled as a cost for iterative deepening. One explanation for this behaviour is as follows: The reasoner commits to the choice of E_2 for transporting the orange juice from Elmira to Corning. Later, the reasoner commits to keeping E_2 in Elmira until this event, since initially E_2 is in Elmira. Thus, E_2 cannot be used for getting oranges, because it is kept in Elmira and E_3 is the next option.

Concerning the major decisions of the reasoning process, we give a brief analysis of the resulting log file (the percentages indicate the time of the respective decision):

- Relatively quickly, the minimal time of 8 hours is detected. [6%]
- Reasoner commits to keeping E_2 in Elmira. [6%]
- It takes long to prove that the oranges cannot move before 3 AM. [57%]
- Boxcar has to be from Elmira. [67%]
- The engine for moving the oranges is E_3. [70%]
- The engine for moving the OJ is from Elmira. [76%]
- Arriving in Elmira with at least two cars. [82%]

The search space of this test case is illustrated in Figure 5.3.

The boxed area indicates the centre of the tableau structure. Each node represents a reasoning state that was created by branching and was subsequently closed.

The analysis of the test case indicates that even a small number of additional constraints (as compared to the enumeration approach discussed in

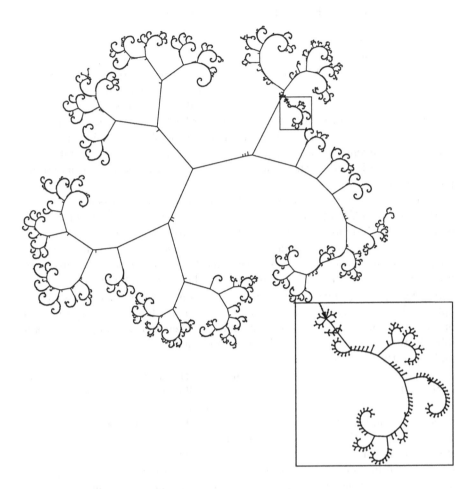

Fig. 5.3. An illustration of the search space in a TRAINS test case.

the last section) can reduce the search space in order to effectively find corresponding solutions to more complex test cases. Both experiments conducted in the TRAINS domain show that in order to be of a practical use, more information to restrict the search space is required. However, they also indicate an interactive approach involving a human user may be a solution. In particular, some of the hypotheses that the reasoning engine decides to adopt and evaluate should be reconfirmed with the user. Given the interactive protocol that CIDRE implements these decisions can be communicated at the time they are taken and a user may interfere effectively.

5.5.5 A Distributed Batch Processing Architecture

In order to run autonomous batch tests with the reasoner and to distribute the work load on different hosts in a network, a simple batch processing architecture has been implemented. The central component of the architecture is a network-accessible relational database. In this database, batch processes, hosts, and other items are described and are updated as necessary at run time. Client programs connect to the database in order to create a new batch job, initiate the processing an existing still unassigned job, and update information on the currently processed job. The clients also update information on the current work load of the host. This information is used when assigning jobs to hosts.

The usage scenario is as follows: One or more job processor clients are started on the different hosts in the network. The number of job processors is not arbitrarily bounded. The job processors are initially in an IDLE state where they periodically connect to the job database and try to obtain a new task to process. One or more jobs are created by a job creator client program. Each job is described by the domain theory to be used, the build version of the reasoning engine, and certain run time parameters such as reasoning strategy options. Furthermore, certain criteria for selecting a host can be specified in order to, for instance, only use a machine with enough memory for certain jobs. In the JOB table an ID value is automatically assigned to the new job and its status is set to OPEN.

One of the idle job processors connects to the database and detects the OPEN job. If the jobs host requirements are fulfilled, it reserves the job, puts it into the PROCESSING state and starts to run the process. Otherwise, it disconnects and retries after a certain delay. Once a processor has reserved a job, it reads its input specification and parameters to be used. Then, it starts the model generation procedure as a subprocess. While this process is running, the job processor periodically connects to the database and updates a variety of run time statistics. If the processor crashes for some reason (e.g. because the host machine is powered down), the database connection is lost and the job is not updated anymore. These kinds of jobs will be spotted by using an administrative tool and can be subsequently restarted. If the job processor terminates successfully, either because one or more models could be generated, or the theory could be proved inconsistent, it assign the job the SUCCESS state and stores relevant information in the database (e.g. concerning the models found). The processor then returns to the IDLE state.

In addition, the batch architecture described here strives for portability, fault tolerance, and extensibility. To this end, the database server and client programs are written as pure Java programs and thus run on a wide range of operating systems. Furthermore, the job processor clients are robust in the sense that they do not stop their client processes if the database server is temporarily unavailable (for instance due to network problems). The messages to the database are buffered and will be sent as soon as the database is available again. Finally, the processing architecture could be used for other batch processes as well. Only a few specific features are implemented for the task of running reasoning processes, but they need not be used by other applications. Using SQL queries, a number of statistics and reports can be generated in a flexible and efficient manner, which is another provision for a more general applicability of the architecture.

5.5.6 Interactive Testing

Automatically testing the interactive protocol is somewhat more difficult than testing the autonomous mode since it relies on the temporal behaviour of the reasoning engine in combination with the entire computing environment. Starting the reasoning engine with a specified test case and waiting until it finishes is a repeatable process that should produce exactly the same results even if the time required varies significantly between different host machines. The interactive protocol, on the other hand, is designed to operate asynchronously, i.e. independently of whether the reasoner has finished or not.

The approach taken here to test the interactive protocol is to send a timed sequence of protocol messages and to observe the behaviour of the reasoner. Tools for interactive testing have been implemented. These can be used from the command shell, for instance.

5.6 Discussion

This section has introduced CIDRE, our implementation of a Common Interactive Domain-level Reasoning Engine. We have distinguished between the core reasoning process and the interaction capabilities which are provided by its interactive protocol. The core process is based on an existing finite model generation procedure, called Positive Unit Hyper-Resolution. The implementation and adaptation of this procedure has been described as an iterative function that defines the transition from one reasoner state to another. Using such an iteration enables us to implement the interactive protocol in a clean and modular fashion. The goal is that an external client component (such as a dialogue manager) can control concurrent reasoning processes in various ways. The protocol described here generalises and enhances earlier approaches to integrate our reasoning engine [87, 88]. On a more abstract level this can be viewed as a way to realise a synchronisation mechanism between the user's requirements and the related system inferences. This will be the foundation on which the approaches described in the following chapters are grounded.

We view CIDRE and its combination with the domain modelling approach described in Chapter 4 as a practical *programming tool* for concrete domain applications, rather than as a vehicle for proving theoretical properties. In principle, CIDRE can be considered an interpreter for a specific language that provides the domain modelling and problem-specific test cases. In this view, it is vital to support a developer that applies this system to understand how the system works. A prerequisite for this is that we attempt to reduce the amount of concepts and heuristics implemented in the system to a minimum. Consequently, we have opted not to implement special behaviours for integral numbers, for instance. As a programming tool, CIDRE's intended usage may be compared to using a relational database management system or a deductive database system [89]. The key difference compared to a relational database system is that CIDRE is able to produce alternative models, while a conventional database can only represent the content of one model at a time.

Concerning the quantitative tests presented in Section 5.5, one may note that there are apparently more efficient methods of solving these reasoning problems, in particular, if they can be approached differently from model generation. For instance, the "computation" of a list of a small length does not take seconds if Prolog's length/2 predicate is used. Nevertheless, in the model generation approach, what is generated is more than a mere description of the structure. If model generation succeeds it has shown that the data are *consistent* with all the axioms that are encoded in the theory. In this process, the entries of the *member*/2 relation are computed, which may be used for very efficient access at a later stage. In this sense, the procedure is similar to a dynamic programming method. Thus, some of the effort is caused by the complexities inherently involved in the model generation process. Of course, one option here would be to remove particular relations like *member*/2 in the tests where they are not the primary focus. However, this would probably reduce the significance of the test results concerning realistic applications based on our library of domain models.

Another factor specific to CIDRE is, however, that on top of computing the resulting models the engine also provides extensive documentation of its reasoning in terms of proof structures. Although some of this information can potentially be used to optimise the reasoning process, it is not required for determining the models as such. Instead, this information is required for the enabling the dialogue management approach we propose. Of course, more research into how these structures may be built more efficiently may provide better quantitative results in the future. However, this has not been the focus of our research.

Concerning the TRAINS domain, in particular, the results so far indicate that there is still a range of options to optimise the current behaviour. One of the strategies that we will briefly mention here is parallelisation. From various test cases we have observed that the reasoner takes to explore alternative hypotheses can vary substantially. Thus, one idea is to try different hypotheses concurrently on different host processors or processor cores. For a more general use of parallelisation, one question that needs to be addressed is how information from parallel reasoners can be efficiently integrated. An aspect of this problem is that there may be significant overlaps in the inferences that each of the parallel reasoning processes will produce. Equivalent inferences should be recognised and represented only in one instance.

6

A Prototype Based on VoiceXML

6.1 Introduction

In this section, we discuss the design and implementation of a prototype that features the integration of our logic-based reasoning approach into a VoiceXML-based dialogue system and architecture. More precisely, a sample domain and dialogue application has been implemented and integrated with the reasoning engine which deals with the domain-level requirements and inferences arising from the interaction.

VoiceXML can be viewed as an attractive platform because of its growing industry acceptance as a standard for voice-based applications. In addition, it is supported by a growing variety of development tools and deployment infrastructure. Thus, an increasing amount of developer software in the form of reference or research implementations is becoming available. This has contributed to a growing base of developers who create applications within the VoiceXML framework. In addition, solutions for deploying these applications to actual users has been significantly simplified due to offerings from different companies specialising in the hosting of voice applications.

Thus, the goals for the VoiceXML prototype are the following ones. First, the prototype should provide insights concerning how the interface between a VoiceXML-based application and our reasoning engine may be realised. In particular, the goal is to build upon the form-based interaction. The benefit thereof consists in the potential to extend existing VoiceXML applications, rather than requiring to start from scratch. A related goal is to determine to what extent our proposed dialogue management approach based on requirements and inferences is compatible with the VoiceXML framework. Finally, the prototype system should enable us to clarify the requirements concerning the linguistic front-end in that context.

In the following, we will describe the application domain selected for the prototype, the approach to dialogue design on the basis of VoiceXML, and the architecture that integrates the VoiceXML dialogue manager with the reasoning engine. We discuss several areas in which building the prototype has led to new insights. In particular, we explain which of the limitations of the VoiceXML approach are the most important to be addressed in a revised architecture.

6.2 VoiceXML Platform

The VoiceXML platform developed at the University of Ulm [90, 91] was chosen as the basis for the prototype. In this platform, a subset of VoiceXML 2.0 has been implemented as a research system. The main difference to most existing implementations of VoiceXML is the fact that the core of our research system is a compiler [92, 93]. It statically preprocesses the VoiceXML application and produces a stand-alone program that performs the dialogue actions when run. Of course, this limits to some extent the flexibility that is offered by VoiceXML. In particular, it prevents the dynamic generation of VoiceXML documents, and therefore the implementation can only be considered a subset of VoiceXML. On the other hand, there are a number of potential advantages of the compilation approach. First of all, the integration of external code is substantially simplified in the compilation approach. In fact, the complete Java class library as well as other class sources can be used. This was the primary reason for using the compilation approach, since the reasoning engine code has to be integrated tightly with the dialogue. In addition, whenever compilation is possible it results in more efficiency and less resource consumption at run time. Since the compiled code does not require a web server for hosting the VoiceXML documents nor an XML parser for analysing them, the run time environment is substantially less complex. In fact, one of the motivations for the compilation approach was to enable the use of VoiceXML-based dialogues in rather restricted environments, such as embedded systems. Finally, from the dialogue developer's perspective, off-line compilation can be a step toward better code analysis and to the detection of certain kinds of errors that would otherwise only be detectable than at run time. Variable references are one example of this property.

The compilation is performed in the following steps:

1. First, the VoiceXML root document, referenced documents, and any referenced grammars are acquired, parsed and translated into an internal representation. As part of that process, EcmaScript code portions [94], for instance, in condition attributes, are also translated. This process detects missing elements such as forms in VoiceXML or missing rules in grammar documents.
2. Based on the internal representation a stand-alone program consisting of EcmaScript code is generated. The usage of EcmaScript as an intermediate format is a straightforward choice, since it is an integral part of the VoiceXML specification and is used in many places in VoiceXML documents [32]. The generated code implements a specialised version of the

VoiceXML Form Interpretation Algorithm in the sense that it implements the Select-Collect-Process cycle for each VoiceXML form in the application. Form items are represented as record members in their respective generated form data structure. One challenge to be noted at this stage is the ensure that EcmaScript code fragments from the VoiceXML document can be used without modification (for instance, the correct variable scoping has to be ensured).

3. The combined EcmaScript program is subsequently compiled into Java byte codes by the Mozilla Rhino JavaScript engine [95]. The result is an executable Java archive consisting of a number of generated classes.

The resulting code accesses a generic interface to speech interaction components, i.e. speech recogniser and prompter engines, each of which features a number of essential interface methods. The methods can be implemented by different speech engines. For instance, for maximum portability, Java implementations like the Sphinx-4 continuous speech recogniser [96] and the FreeTTS speech synthesis system [97] have been used. On the other hand, the Java Native Interface (JNI) provides the means to access other implementations available in native code.

6.3 Accessing the Reasoning Engine from VoiceXML

In the following, we describe some technical aspects of how a dialogue manager implemented on the basis of the platform sketched above can access our reasoning engine CIDRE in order to exchange requirements and inferences. The main task of the interactive protocol (cf. Section 5.4) is to translate between the Java-based functionality accessed from within the VoiceXML dialogue and the reasoning engine which is implemented in the Haskell programming language. Thus the text-based communication channel which is offered by the interactive protocol was chosen for the integration. The following general core functions of the interactive protocol are the most important: the creation and deletion of reasoning tasks, based on provided sets of assumptions and flexible polling of new reasoner messages under certain filtering criteria. A Java layer that encapsulates the interactive protocol manages this communication on the Java side, such that the Java layer can be used with regular methods and data structures. The Java layer provides a complete object-oriented implementation of the base data structures used in the reasoner, such as recursive proof structures, as well as clauses, atoms, and terms.

In addition, the Java layer provides an auxiliary class SlotReasoner as a wrapper around the reasoning engine. Instances of this class maintain a current set of requirements, based on the notion of slots. We use the term *slot* to refer a parameter that may be assigned a value to in the dialogue. When the requirements are modified, this class internally uses the reasoner's interactive protocol to update the reasoning process.

For the purpose of the description, the most important methods the `SlotReasoner` class provides are the following:

`clearSlot(slot)`, `addSlot(slot,predicate)`, `addSlotNeg(slot,predicate)`:

> These methods may be used to drop or create requirements for a slot that is identified by its name. An updated reasoning process for the new set of requirements will be created by the `SlotReasoner` class.

`hasConflict()`, `hasSolutions()`, `getSolutionCount()`:

> These methods may be used to determine whether the reasoning process has resulted in a conflict or generated any solutions.

`hasInferredSlot(slot)`, `getInferredSlot(slot)`:

> In the case the reasoning produced a single solution, these methods can be used to derive respective value of the given slot.

The basic behaviour of a VoiceXML dialogue is governed by the Form Interpretation Algorithm (cf. Section 2.5). The integration of VoiceXML-based dialogues with the asynchronous reasoning engine has to provide a bi-directional communication mechanism. In this communication mechanism, one direction, namely the communication from the dialogue to the reasoner, is rather straightforward because form item values and updates thereof can be conveniently mapped to operations provided by the reasoner's interactive protocol. In particular, form handlers provided by VoiceXML can be implemented to handle these kinds of situations within the conventional FIA processing cycle.

The reverse direction, i.e. the communication from the reasoner to the dialogue manager, is much more complicated. In particular, the VoiceXML FIA is not designed to be easily interruptible, since it assumes to be in control of the interaction. An interruption of the FIA, however, especially while waiting for user input, may be desirable when the reasoning engine produces an important inference that influences the dialogue. Consider the following example:

> U-1 > Find a Mongolian restaurant.
> S-2 > [*communicates requirements to reasoner*]
> S-3 > Ok.
> S-4 > Which location?
> S-5 > [*reasoner detects conflict*]
> U-6 > Centre.
> S-7 > Sorry, there is a conflict.

Here, the subdialogue S-5,U-6 becomes irrelevant, once the reasoner information from S-5 becomes available. However, as such, the dialogue is not in control anymore once it waits for input (S-4). Two strategies for tackling this problem seem possible. The first one consists of polling the reasoner state in regular intervals, i.e. by limiting the time spent waiting for user input by using short time-out durations. After a time-out, the reasoner can be queried, and the FIA may continue to wait for user input unless the reasoner has provided new inferences. In our current prototype we have opted to implement this strategy.

Alternatively, and possibly more elegantly, an event-based method seems possible. In this method an event would interrupt the FIA if it is currently waiting for user input. The event would need to be generated by the VoiceXML platform in response to an event obtained from the reasoner. The event would enable to dialogue manager to regain control and respond to the new situation. This would potentially result in fewer implications on the dialogue design, since polling may be avoided. A potential problem may arise here since not all inferences are equally important, i.e. not all inferences should be allowed to interrupt the current dialogue flow.

Technically, a third possibility exists in which input from the reasoner is processed in the same way as user input. This means that an inference obtained through the reasoner's interactive protocol would have to be wrapped into a semantic structure. This semantic structure would need to include special key-value pairs, such that when the FIA processes it, the dialogue script has appropriate handlers installed. However, both this method and the event-based method described before would require the modification of part of the VoiceXML platform. This is possible for a research system, but limits the portability of the application.

6.4 Application Domain

The domain of the sample application is a restaurant selection (cf. Section 4.4.8) based on several parameters (slots). Although this can be achieved in a rather straightforward way with a more conventional relational database approach, the reasoning-based method arguably provides certain advantages in terms of flexibility and scalability to further extensions. For instance, using our proposed reasoning engine to maintain requirements such as the general location of the restaurant may enable an extended system to provide travel information. A dialogue in the sample restaurant domain consists of exchanges in which the user specifies parameter values to restrict the list of possible restaurant options. The system reacts by presenting a summary of the options available or by presenting information about conflicts that prevent a solution. In the case of a conflict (i.e. an overconstrained situation), the information provided by the system should enable the user to retract or relax some of the constraints he introduced.

The main search parameters concerning restaurants are:

Location: The values of this parameter are organised in a spatial tree-structured hierarchy, e.g. the city area is subdivided into districts.

Cuisine: In a similar way, the values of this parameter are also structured from abstract categories into more detailed ones. For instance, the concept of

Asian cuisine is a higher-level category that comprises, among others, South-East Asian cuisine, which in turn comprises "concrete" concepts, such Indonesian cuisine.

Name: Using this parameter, the restaurant can usually be identified directly, with the exception of fast food chains.

The goal of the value hierarchies is to enable the user to first specify requirements on a rather general level. This may be desirable for novice users, in particular, who may not be aware of the "concrete" categories available. In addition, the higher-level requirements concerning location and cuisine can be combined and lead to detailed solutions.

6.5 Dialogue Design

This section describes the design of the example dialogue application. As a VoiceXML dialogue (cf. Section 2.5), the design is based on the form-filling style of interaction, but enhanced with the reasoning-oriented aspects. In this prototype, however, the reasoning engine's capabilities are not yet fully taken advantage of, and suggestions to further enhance the dialogue design in order to do so are presented in Section 6.6.

In the form-filling approach, the information necessary to perform a dialogue task is usually divided into certain smaller and simpler units. These are represented in VoiceXML as fields in a form. A field can store an arbitrary semantic object as its value. Usually the value is obtained from the language interface. However, fields can also receive values via assignments from within the form, for instance, in the case of inferences such as calculations or database look-ups. A field is a chunk of information that can be individually asked for by the system. Its value can also be filled by an overanswering utterance by the user (i.e. an utterance that contains multiple fields values.)

In our example application the concepts mentioned above are reflected in the following way: There is only one task, namely to identify a restaurant in a database which matches the requirements introduced by the user. This task is represented by a single VoiceXML form. The form has fields for the following value constraints: the requested location, the cuisine, and the name of the restaurant. A value x of field "cuisine", for instance, represents constraint that $cuisine(r) = x$ for concrete cuisine values x, where $cuisine(r)$ is the cuisine of the restaurant to be found. For abstract values, the constraint would be stated in the more general form $x(cuisine(r))$, which allows, for instance, $asian(cuisine(r))$ or the more concrete constraint $southeastasian(cuisine(r))$. For each parameter field, there is an associated field that is used for confirming the value if the value was provided by the user or for indicating the negotiation status of the value if it was set by a system inference. The "name" field is intended for system output in most cases, because it is assumed that in the most frequent cases the user will specify the location or cuisine and the system is to present suggestions including name. Nevertheless, the user might also use the name parameter as in "Where can I find a Starbucks?"

6.5.1 Form Structure

In this section, we describe the form structure used to implement the dialogue. The dialogue script of the prototype consists of only one form. The form is used for storing the slot values and for communicating inferences to the user. The structure of the form is intended to be an extension of the structure commonly found in VoiceXML applications. For instance, a form-level grammar is used to allow overanswering.

The reasoner, more precisely, an instance of the wrapper class SlotReasoner, is represented in a form variable R. The form contains some basic scripts and blocks to communicate with this instance (Lines 5 to 11). In particular, these blocks handle the cases of conflicts or solutions.

For each slot S (i.e., location, cuisine, name) the following fields are contained in the form, cf. Figure 6.1.

fi$S$ is a conventional field for representing the slot value. This may be from an existing application. The new aspect is that there is a <filled> element that communicates the value to the reasoning engine. Any previous value is overwritten. This also clears any previous negated values.

fi$S$Neg is used to allow the specification of negative requirements. Note that in multiple utterances, different negated slot values can be collected. These are added to the reasoner. This slot is not intended to be used for prompting.

inf$S$ is a field that is used to reconfirm an inferred value. Depending on the user answer is either copied to fiSlot, or it is negated. In the latter case, however, a conflict will arise.

For the sake of readability, a couple of aspects have been omitted from the code fragment. First of all, the handling of asynchronous reasoning is not shown. Second, some more assignments have to be made to ensure the correct order of blocks and items. In particular, to enable the conflict handling and summarisation of results. Finally, the actual prototype was in German, so the prompts and grammars are presented in a translation here.

Like an ordinary VoiceXML application, the prototype uses both form-level and field-level grammars. The tasks of the grammars are to allow the user

- to provide slot values,
- to provide negated slot values ("not S"),
- to clear a slot ("drop S"),
- to direct the focus to a slot ("what about S"), and
- to issue basic information requests ("how many", "which S").

```
 1    <form id="main">
 2      <var name="R" expr=new Packages.SlotReasoner()/>
 3      <grammar src="grammar.srgs"/>
 4
 5      <block name="conflict" cond=R.hasConflict()>
 6        Sorry, there is a conflict.
 7        ...
 8      </block>
 9      <block name="summary" cond=R.hasSolutions()>
10        There are <value expr=R.getSolutionCount()/> restaurants.
11      </block>
12      ...
13
14      <field name="infS" cond=R.hasInferredSlot('S')>
15        <prompt> So, S is
16          <value expr=R.getInferredSlot('S')/>? </prompt>
17        <filled><if cond="infS">
18          <assign name="fiS" expr=R.getInferredSlot('S')/>
19          ...
20        <else/> R.addSlotNeg('S',fiSNeg); </...>
21
22      <field name="fiS">
23        <grammar uri="S.srgs"/>
24        <prompt> Which S? </prompt>
25        <filled> R.clearSlot('S'); R.addSlot('S',fiS); </...>
26
27      <field name="fiSNeg" cond="false">
28        <filled> R.addSlotNeg('S',fiSNeg); </...>
29        ...
30    </form>
```

Fig. 6.1. VoiceXML fragment for reasoning-enabled forms. Scripted code is written in type writer font. The elements containing S are repeated for each slot.

Domain-specific grammars have been designed for the application domain. Field-level grammars have been designed for the fiS fields. They simply allow to provide a respective value for the field. In the application domain, the slot values can be uniquely mapped to their respective slot. Thus, the form-level grammar is essentially a phrase spotting grammar. Parts of the grammars have been generated from the basic classification descriptions (cf. Section 4.2).

6.5.2 A Sample Interaction

This section discusses and analyses a sample interaction with the prototype. A rendition of an interaction slightly glossed for readability is given in Figure 6.2.

Utterances S-1 and S-2 form a system-initiated opening by greeting the user and asking an initial question. The user responds in U-3 with a high-level

```
S-1 > Hello!
  2 > What kind of restaurant do you want to find?
U-3 > I want to eat Asian food.
S-4 > There are 17 restaurants.
  5 > What is the location?
U-6 > Bengali.
S-7 > So, the cuisine is Bengali?
U-8 > Yes.
S-9 > There is one restaurant. The name is Anondo
U-10 > This is in the Weststadt, right?
S-11 > Sorry, there is a conflict.
  12 > So, the location is Centre?
U-13 > Drop cuisine, then.
S-14 > OK.
  15 > There are 3 restaurants.
```

Fig. 6.2. Example dialogue in the restaurants domain.

requirement. The user requirement is translated into a logical form which is then passed to the reasoning engine as a new reasoning task. The reasoning engine is able to finish the reasoning task quickly, resulting in a large number of possible solutions. The system initiates a reaction by summarising the solutions on a rather coarse level, and prompting the user to provide more information in utterance S-5. In U-6, the user narrows down his requirement which essentially identifies a restaurant option in the solution space. This is detected by the reasoning engine and the information can be used by the dialogue manager. Since there is only one option left, the system attempts to confirm the user provided parameters and the remaining parameters which can now be inferred. However, the user assumes the turn in utterance U-10 and tries to confirm his assumption about the location of the restaurant which is in conflict with the information that is available to the system. Technically, U-10 is interpreted as a new requirement which is again passed to the reasoning engine as an extension of the reasoning task before. The reasoning engine detects the inconsistency and notes that it is caused by the requirements of utterances U-6 and U-10. So, the user decides to drop the former requirement and the dialogue continues with this modified set of requirements.

Apart from regular form-filling behaviour, such as confirmation steps, the dialogue is designed to be flexible enough to allow the following reasoning-related dialogue functionalities: The interaction provides a means for the disambiguation of abstract values, such as "Asian cuisine", and their resolution into primitive values such as "Indonesian" cuisine. Furthermore, it is possible for the user to retract a value assignment, in order to drop that requirement from the search without replacing it with another value. The user can also direct the focus of the dialogue to a chosen parameter, i.e. field.

```
U > What about the cuisine?
```

Flexible formulations of complex requirements are possible using the logic-based approach, i.e. the user can state negated value and combinations of values. For instance:

`U > European, but not italian.`

6.6 Discussion

In this section we summarise and discuss some experiences we made and different aspects of using VoiceXML as the basis of a reasoning-enabled dialogue manager.

First of all, we have shown that it is possible to technically integrate a VoiceXML dialogue manager with the reasoning engine CIDRE. The reasoner's interactive protocol is accessed via a software layer written in Java. This layer, in turn, is accessed using VoiceXML's scripting facilities. A more standards-conforming technique to achieve this communication may be the use of VoiceXML's `<data>` element. Using this element, a dialogue can communicate with a server via HTTP. However, we think that, for the purposes of this prototype, the benefits would have been minor compared to the increased effort necessary to realise a server-based communication.

From the user's perspective, compared to a basic form-filling dialogue the integration provides the following enhancements: First, it is possible to flexibly make use of more abstract values which are defined in classification hierarchy within the domain theory. Second, it is possible to use negation to rule out some of these values. Finally, specific inferences, such as the value of a dependent slot or a conflict, are handled by either reconfirming the inference or attempting to resolve the conflict.

To a dialogue developer, the advantages may be the following: First, an existing form-filling application may be reused and enhanced without imposing a completely new dialogue design. Second, the processing of domain-level reasoning can be separated from the implementation of the dialogue strategy. On the conceptual level, this is made possible by interpreting field values in the more general form of user requirements and inferences.

The prototype presents a lot of options for improvements. Concerning the dialogue design, for instance, conceivable enhancements of the approach include a better explanation of inferences and a better conflict resolution. This may be accomplished by inspecting the provided proof structure. On this basis, the fields involved in an inference can be determined. In terms of technology, the automatic generation of parts of the VoiceXML form and the associated grammar structures has already been investigated and may provide a possibility to reduce the implementation effort.

The more fundamental problem is on the conceptual level. Generally speaking, there is no clear way how to take substantially more advantage of the reasoner. Forms, at least when using as key-value assignments for a database search, provide little support for more complex requirements and inferences. For instance, possibilities for the representation of disjunctive requirements seem to be missing. Another problem is that not all inferences map conveniently to dialogue states in the VoiceXML sense. For instance, conflicts often affect multiple slots, not just one value alone.

Further features that may be supported by the reasoner, but are difficult to realise within VoiceXML, are: The user should be able to create "scenarios" which then can be evaluated, compared against each other, modified, or discarded. In addition, it should be possible for the user to explore alternative suggestions by the system.

To summarise, we can conclude that the data model provided by VoiceXML is much more restricted than what the reasoning interface would have to offer (i.e. key-value pairs vs. logical formulas involving logic operators). In addition, tightly integrating VoiceXML with an asynchronous reasoning process is non-trivial.

7

Information State-Based Dialogue Management

7.1 Introduction

The construction of the VoiceXML-based prototype (cf. Chapter 6) illustrates some of the benefits, but also some of the drawbacks of relying on the concepts that a dialogue scripting language provides. Some problems of the integration in that prototype are technical, for instance, the need to interrupt the form interpretation algorithm. Other problems are more fundamental. In particular, the form-based representation of a dialogue does not directly provide an adequate representation of some aspects of our requirements-inferences dialogue management approach. As a consequence this prototype does not take full advantage of the capabilities the reasoning engine offers.

This chapter describes an alternative approach to dialogue management to better exploit these facilities. In particular, we claim that important domain-independent dialogue functions can be implemented on the basis of a relatively small set of dialogue moves.

This chapter is structured as follows: In Section 7.2, the general approach and model of dialogue management based on requirements and inferences is described in more detail. On this basis we develop a generic system strategy to achieve the desired dialogue functionalities. We formalise the system strategy by defining a set of dialogue moves and an information state structure. In Section 7.3 and the following, we describe how different dialogue functionalities can be realised using these dialogue moves. This concerns the integration of user moves into the dialogue state as well as the generation of system moves in response. In particular, we define the role of the reasoning engine, i.e. how dialogue moves interact with the reasoning engine. Finally, in Section 7.7, we discuss certain design decisions and alternatives, as well as possible extensions of the framework.

This chapter focuses on the domain-independent aspects, their application to specific application domain is discussed in Chapter 8. In particular, the approach presented in this chapter is based on some simplifying assumptions. For instance, we assume a relatively powerful linguistic front-end that provides an interface based on dialogue moves. The description of the prototype will present more details as to in how far these assumptions can be realised.

7.2 Collaborating on Requirements and Inferences

This section describes our general dialogue management approach. We begin by elaborating on the terminology established in the introduction chapter. This will provide a useful basis for the description of the dialogue approach.

- A (user) *requirement* is an information or condition introduced by the user during the dialogue. It can be thought of as a goal, constraint, assumption, or belief that the user wants to fulfil or take into account. For instance, such a requirement may be "arrive at 10 AM in X". The condition may not be desired by the user, and may be committed to only temporarily, as in "What if there is a traffic jam?"
- An *inference* is a proof produced by the reasoning engine on the basis of the requirements from the user and potentially its own internal knowledge base. For instance, the internal knowledge base may state that going from Y to X takes at least 2 hours. The difference between requirements and inferences is that requirements are negotiable (i.e. the user can opt to retract them). An inference, on the other hand, is compulsory in the sense that it cannot be retracted while keeping the requirements it is based on because it follows from the domain theory.
- A particular case of inference is a *conflict* which proves the inconsistency of the user requirements. The requirements may be inconsistent on their own or inconsistent when combined with the internal knowledge base of the system. Importantly, in many cases a proper subset of the requirements and the system's knowledge base are conflicting and it is important to determine exactly which requirements are part of this subset.
- A *suggestion* is a statement that the system assumes hypothetically because of some inference rule that states alternative possible postconditions. For instance, given any two non-identical appointments a_1 and a_2, and a rule that states that these may not overlap, without additional information. The system may produce two (alternative) suggestions, one stating the a_1 precedes a_2 and the other one stating that a_2 precedes a_1.
- A *solution* is a situation where the user requirements can be combined with the system's knowledge base and yield one or more consistent models. A solution is an extension of the user requirements with a set of system suggestions. The system suggestions guarantee that all the necessary inferences have been produced. In the terminology of logics, a solution is a saturated model that satisfies all the user constraints. A *potential solution* is a situation where, during the reasoning process, it is not yet clear if a current set of requirements and system suggestions will be consistent or not. We will also use the term *scenario* to refer to solutions and potential solutions.

In this section we describe general assumptions about the kinds of tasks the user and system can solve in collaboration. The dialogue systems we focus on share the following main objective: To allow the user to specify domain-specific requirements and to collaboratively construct a consistent solution that incorporates as many of these requirements as possible. Consistency implies that the domain-specific knowledge of the system (i.e. domain theory) is taken into account and may lead to inferences that the user needs to accept or revise his requirements. In this sense, all the user and system need to do is to collaboratively reason about certain user-initiated requirements. Thus, the general approach to dialogue management is to enable the user to provide information and have the system provide inferences and suggestions based on that information in order to build a consistent and unambiguous solution. In our model the process of building a solution is a collaborative user-system interaction. It consists of turns providing requirements and revising them, if necessary, upon evaluation by the system. Concerning the roles in the dialogue, the user is assumed to provide high-level information. The system, on the other hand, provides specific data and details. The user provides requirements, and the system provides inferences based on these. The dialogue management has to decide how to organise this interaction.

In some systems, such as Smith et al.'s Circuit Fix-It Shop, building a solution is viewed as the construction of a proof of certain goal statements. However, our approach is based on the notion of consistency, rather than on the notion of goals. Our approach is also to some extent similar to the artificial discourse language for collaborative negotiation proposed by [29]. The key difference is that their model is based on the construction of problem-solving plans whereas our model focuses on the domain level.

We do not consider primarily action-oriented domains where the main goal of the system is to immediately interact with the external world. Such systems are used, for instance, for controlling a VCR or commanding a robot to perform some action. However, in our approach the user and system may collaboratively *reason* about the possible actions and their consequences. In this sense, our dialogue management approach may complement these action-oriented systems.

7.2.1 A Generic System Strategy

The aim of this section is to describe a generic system strategy that achieves the desired properties and functionalities of our requirements-inferences dialogue management approach. The basic algorithm in Figure 7.1 illustrates the proposed strategy. This algorithm will be repeated on any event as long as the system is running.

In the algorithm the main steps can be described as follows: The user should always be able to retract a requirement, either because it is part of a conflict or for other reasons (Lines 1 and 2). In this case the system accepts the retraction and returns from the processing.

Algorithm systemStrategy

```
 1: if user retracts requirement then
 2:     accept
 3: else if conflict then
 4:     suggest resolution
 5: else if user asserts requirement then
 6:     accept
 7: else if there is a user question then
 8:     generate answer
 9: else if important inference then
10:     reconfirm
11: else if multiple solutions then
12:     generate question
13: else
14:     report success
15: end if
```

Fig. 7.1. Basic algorithm for determining the system actions.

If the current situation implies a conflict, i.e. a logical inconsistency in the user requirements (Line 3), the system attempts to propose a solution (Line 4). Such a proposal consists of an explanation of the conflict, including the requirements that are inconsistent. In addition it may contain a proposal to the user to retract one of these requirements in order to relax the overconstrained situation. Conflict resolution will be described in greater detail in Section 7.5.1.

If there is no conflict, then the system will accept any new requirements from the user (Lines 5 and 6). Section 7.3 will provide more details on how this is performed in collaboration with the reasoning engine. Otherwise, if there are no new requirements, the algorithm proceeds to Line 7. If there is a current user question, i.e. a question has been raised by the user and has not been answered yet, the system attempts to generate answers (Line 8). In terms of logics, the system views a user question as a request of a proof of a certain statement. An answer may either be a proof of the statement or a signal that no proofs can be found. In both cases the question is updated in order to take the answer into account. The kind of processing necessary to do this will be described in Section 7.4

If there are important inferences that need to be communicated, the system will attempt to reconfirm these with the user (Line 10). This includes an explanation and thus is similar to the handling of conflicts. In fact, a conflict is a special case of inference. However, the conflicts are treated at a higher priority because they prevent further reasoning. Non-conflict inferences, on the other hand, are only communicated in case there are no new user requirements or open questions. The question what constitutes an important inference may be a domain-specific decision.

In Line 11, the algorithm determines whether an underconstrained situation exists, i.e. whether there are multiple solutions arising from the current user requirements. In that case, the system will attempt to resolve the ambiguity by generating a related question to the user. There may be multiple strategies for selecting a system question. For instance, the current context should be taken into account. In addition, this processing may also be performed in the case the reasoner has not yet finished. In this case, the question may attempt to suggest one of the models already found or to gather new requirements from the user. The latter case serves to restrict the reasoner's search space.

In the remaining case (Line 14), a unique solution has been found for the requirements introduced by the user. This means that, as far as the system is concerned, the dialogue has been successfully completed. The system therefore may ask whether there are any other requirements that need to be taken into account ("What (else) can I do for you?"). In a typical scenario, this will also be the initial situation in the dialogue, since we assume that initially the set of user requirements is empty.

Overall, the generic system strategy may lead to situations where requests are accepted by the system even though they lead to conflicts with existing information. In general, this is not avoidable since the conflict detection may take some time and thus it is not clear if a request may be rejected immediately. Even if the conflict can be determined quickly, the system should offer to the user the option to replace conflicting earlier requests with the later one.

7.2.2 Dialogue Moves

We have chosen an information state-based approach as the basis for our implementation of the generic system strategy introduced in the last section. Thus, the dialogue manager in our approach is a dialogue move engine, i.e. it uses dialogue moves to update its internal information state structure. Defining dialogue moves allows us to give a better characterisation of the processing steps involved in the generic strategy. However, we follow the structure and syntax of update rules and update algorithms proposed by e.g. TrindiKit [39] and related systems only on a rather conceptual level.

On an abstract level an information state-based approach consists of the following components:

- a set of dialogue moves,
- an information state structure,
- an update algorithm that explains how user dialogue moves are integrated and how system moves are generated is response.

We describe these first two components in the following. The update mechanisms will be described in Section 7.3 and the following. Table 7.1 shows the domain-independent dialogue moves that both the user and system can use for the interaction. Essentially, these moves can be thought of as a domain and language-independent abstraction of actual utterances.

Category	Move	Description
Conventional	Hello	Greeting
	Bye	Closing
	Ok	Success or opening the conversation
Requirements/	Assert x ϕ	Adding a requirement or inference
Inferences	Retract ϕ	Removing one
Requests	Query x ϕ	Requesting information
	Done	Finishing the response
Hypothetical	Enter ϕ	Adding a hypothesis
	Commit	Accept hypothetical results
	Cancel	Reject results
Focus	To ref	Focus on an entity
	Other	Switch focus

Table 7.1. The dialogue moves in the information state-based approach.

We distinguish several high-level categories of dialogue moves. Firstly, *conventional moves* initiate and terminate the interaction. These moves will not be discussed in detail here. The *requirements* category concerns the management of requirements and inferences, i.e. providing or retracting information. Providing information can either refer to the user stating some requirement or to the system presenting an inference or suggestion based on the current set of requirements. The user can also retract requirements using the Retract dialogue move. The category *Requests* is used for requesting information. It includes asking for information and both Yes-No and Wh-questions. In order to enable hypothetical reasoning, the user can employ dialogue moves to enter, and leave the so-called hypothetical mode of interaction. By entering, a new context is created in which requirements can be stated hypothetically, i.e. these requirements can be easily disposed of when necessary. The requirements can also be committed, i.e. accepted as regular requirements. Finally, a *Focus* category of dialogue moves concerns the management of the focus of interaction. This is important for organising the dialogue, i.e. to be able to switch focus between different entities or solutions.

7.2.3 Information State

The structure of the information state used in our approach is illustrated in Figure 7.2:

/SHARED/LU is a list of dialogue moves performed by the user and the system. Each of the moves is described by a feature structure. We describe the details in Section 7.3 and following. In particular, when the user has produced

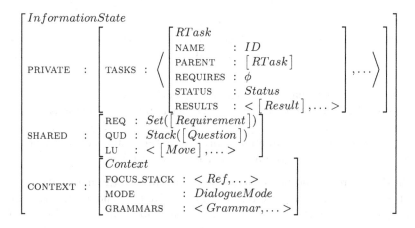

Fig. 7.2. Information state for our dialogue management approach.

an utterance, the respective dialogue move is the first element of this updated list. This information is interpreted by the control algorithm (to be described in the following paragraphs) in order to determine the appropriate system reaction.

/SHARED/REQ contains the requirements introduced by the user.

/SHARED/QUD contains questions raised by the user or system.

/PRIVATE/TASKS is a list of structures representing reasoning task (*RTask*). It contains both the currently running and the finished reasoning tasks. The reasoning tasks are initiated by the system and updated as soon as new information from the reasoning engine is provided through the interactive protocol. Each reasoning task is structured as follows:

- NAME: an identification used to refer to the specific reasoning task.
- PARENT: a reference to the task.
- REQUIRES: a set of requirements that the reasoning is based upon, relative to the parent task.
- STATUS: an attribute representing the current state of the reasoning task, i.e. if it has been finished or is still running.
- RESULTS: a list containing information on each result, i.e. indicating if the number of (positive or negative) results, and the proofs produced by the reasoner.

The list can also be regarded as a stack, in which the top-level element represents the current reasoning task that the dialogue operates on. The current reasoning task represents (what the dialogue manager beliefs to be) the common ground with the user.

/CONTEXT is a representation of the current linguistic context that is used for interpreting and generating natural language. This structure contains information that is mostly relevant to the linguistic interface components. It is described here for completeness and to understand the revised prototype system (cf. Section 8.6). Certain dialogue moves update the context, for instance, to adapt the focus to a newly introduced object. The discourse context itself is structured as follows:

- FOCUS: a stack of entities which are currently in focus.
- MODE: the dialogue modality, i.e. whether hypothetical mode is activated or not.
- GRAMMARS: a list of grammars representing the systems expectations concerning the next user input (contextual interpretation).

In the figure, we do not include the static parts of the information state, i.e. those parts that are always left unchanged by dialogue moves. These static parts are included in the so-called *total information state*. In our case, this mostly concerns additional domain-specific information, which will be described in the discussion of the prototype in Chapter 8.

Also, one may note that the structure of our information state is quite different from the structure used in other approaches, e.g. [31]. This is due to the fact that we want to concentrate on modelling how dialogue moves are realised using the reasoning engine. We are not, to the same extent, concerned with other issues commonly associated with information state approaches, such as, for instance, *grounding*. For modelling grounding the main structural decision is to separate a private from a shared portion of the structure. Although parts of our approach may be cast as a form of grounding behaviour, for the sake of understandability we do not go into details in that direction. Nevertheless, we argue that our approach may complement these systems with the domain-level processing we propose.

Concerning the actual implementation, we view an information state essentially as a typed feature structure [38], i.e. static data type information is associated with each part of the information state. Lists are written in an $< x_1, x_2, \dots >$ notation, which may be considered syntactic sugar for regular feature structures. List structures are also used as stacks.

At the beginning of an interaction, the information state is essentially empty, i.e. it does not contain any dialogue moves or reasoning tasks. In this situation the system will generate an opening dialogue move.

To summarise, we note that except for conventional and the focus moves, most moves deal with asserting and retracting domain-level information, i.e. the dialogue manages the exchange of domain-level information between the user and the system's reasoning engine. In the next section, the individual dialogue moves will be described in more detail. In particular, we discuss how to realise certain dialogue moves on the basis of the reasoner's interactive protocol.

7.3 Processing User Requirements

Processing user requirements concerns the introduction of new information by the user, as well as their retraction. The user can state requirements using the (**Assert** x ϕ) dialogue move, where ϕ is a logical formula. For instance, consider the following utterance:

> U > Add another appointment.
>   (**Assert** x $\{appointment(x)\}$)

When the system processes a user (**Assert** x ϕ) move, it is integrated into the current information state in the following steps:

1. The requirement is added to /SHARED/REQ.
2. A new reasoning task including the requirements ϕ asserted is initiated based on the current one. This is performed by sending a **CREATE_TASK** message to the reasoner (cf. Section 5.4.), where the current reasoning task is used as a parent. This message triggers the task's asynchronous processing through reasoning engine. If the task does already exist, for instance, due to an earlier requirement, it can be reused. The information state is updated to include the new reasoning task as the current one. In addition, the discourse context may be adapted to include the referenced entity as the new focus.
3. A system obligation to address the assertion is implicitly represented in the information state. The assertion is addressed by presenting related reasoning results as soon as they become available.
4. If the reasoning task created in the processing of this requirement results in inferences that resolve open questions, these are removed from /SHARED/QUD.

The **Assert** move provides a generic way to assert a new requirement. In particular, the possibility to assert arbitrary logical clauses makes adding user requirements very flexible. Translating the natural language utterances into this logical form is the task of the linguistic interface, which we will not describe here. However, we discuss some possible natural language utterances and how they can be interpreted as assertions. The user should be enabled to specify complex requirements, i.e. to use logical operators such as negation, disjunctions, conjunctions in order to form complex requirements out of simpler ones.

In the process of constructing the solution, the user will want to provide the information in a flexible and incremental way. For instance, the user should be allowed to provide the information in the order that is most suitable for him without having a certain order imposed on him by the system through a rigid dialogue control.

For instance, the user may introduce requirements by utterances like the following ones:

U-a > I want to eat Indonesian food.
 (Assert $\{$"indonesian" $= cuisine(r)\})$
U-b > I want to eat Asian food.
 (Assert $\{asian(cuisine(r))\})$
U-c > I want to travel to Europe this summer.
 (Assert $\{travel(e),\ europe(to(e)),\ summer(when(e))\})$
U-d > I want to send two boxcars of oranges to Bath by 7 A.M.
 (Assert $\{send(e),$ "Bath" $= to(e),\ timepoint\_precedes(end(e), 7)\})$
U-e > I don't want to eat French food.
 (Assert $\{french(cuisine(r)) \rightarrow \bot\})$

A requirement may also be to some extent abstract or vague in the sense that it may require some disambiguation in order to be fully interpreted. This is useful for allowing the user to describe the problem at first on a rather broad and general level of abstraction, leaving room for different solutions to be found by the system. For instance, in utterance U-a "Indonesian" may refer to an elementary class, whereas in utterance U-b, an abstract superclass is used. Depending on the application domain, the same approach may also be used, for instance, for geographical and temporal expressions (cf. utterance U-c).

A single utterance may include requirements concerning multiple individual task parameter values as in utterances U-c and U-d. This corresponds to the possibility of overanswering in a mixed-initiative dialogue. For the sake of readability we use a set notation in these examples, but the correct processing will be based on individual Assert requirements.

In addition, we do not disallow complex formulas, even involving quantification, but in the most common interactions – and also due to the capabilities of the linguistic interface – conjunctions of ground literals are used. The most basic form of complex requirements are negated requirements (cf. utterance U-e). The use of negation is useful in particular to reject system proposals. More complex formulas may require some preprocessing to translate them to the clause format expected by the reasoning engine.

One situation has to be treated in a particular manner, namely, the creation of *new* entities from assertions. The current convention is as follows: the variable term y in the logical clauses ϕ denotes the entity that is in the current focus. The variable term x is used to refer to an entity that is to be created as a new, distinct entity (i.e. different from the "known" ones). To this end, the dialogue engine maintains a counter that is increased whenever a new entity is created.

Similarly to introducing requirements, the user can also retract requirements using the Retract move. In that case the corresponding requirements are removed from the current reasoning task, i.e. this is different from negating a requirement. Given a current task with requirements ϕ, a (Retract ψ) move is processed by the system in the following way: First, the requirement is removed from /SHARED/REQ. Then, a new baseline reasoning task that does not contain requirements from ψ, but includes as many other requirements from $\phi \backslash \psi$ as possible. This may or may not be one of the ancestors of the current task. If there are multiple options, the system chooses one. Since the initial reasoning task has an empty set of requirements, this can be used as a fallback. Next, this baseline reasoning task is extended by asserting the missing requirements, if necessary. A related problem is the retracting of inferences by the system, which is discussed in Section 7.5.

7.4 Answering User Questions

User questions can either be Yes-No questions or Wh-questions. Yes-No questions should be answered by a simple affirmative or negative response. Wh-questions, on the other hand, describe some entities and should be answered by the system with an enumeration of matching entities. For the purposes of our approach, both types of questions may be handled by the same processing.

The user can state questions using the (Query x ϕ) move where ϕ is a conjunction of atomic formulas, and x is a variable that is used to refer to the described entities.

Consider the following dialogue example in the calendar domain:

U-1 > Do I have any appointments?
    (Query x $\{appointment(x)\}$)
S-2 > You have one appointment at eight until nine
    (Assert $\{appointment(a_1)\}$)
    > and another one at eleven.
    (Assert $\{appointment(a_2)\}$)
    (Done)

U-3 > What's the duration of this meeting?
    (Query x $\{duration(y, x)\}$)
S-4 > Two hours.
    (Assert $\{duration(a_2, 2)\}$)
    (Done)

In utterances U-1 and S-2, a user query for entities is processed, whereas in utterances U-3 and S-4, the user query focuses on a specific property of the current entity. In addition the user may repeatedly request more information by an utterance like "any other?". The system reacts by asserting information that matches the user's query, in particular, two known instances that match the description $appointment(x)$ are communicated. The final Done move signals that the system cannot produce any other answers.

We describe how these kinds of interactions can be generated in the following. Essentially, the system produces a proof for the existential statement

$$\exists x \ . \ appointment(x)$$

by adding its negation to the current set of requirements and then iteratively relaxing the constraint when a conflict can be proved. This is analogous to the proof-by-contradiction mechanism used in Prolog, for instance. To be more precise, a (**Query** x ϕ) move is integrated into the information state in the following steps:

1. A question structure (essentially representing the move) is added to /SHARED/QUD.
2. A new reasoning task including the negation of the conjunction $\forall x . \phi \rightarrow \bot$ is created, based on the current one by issuing a message of the form **CREATE_TASK** $\{appointment(x) \rightarrow \bot\}$ to the reasoner.
3. As soon as the new reasoning task finishes, it is inspected in order to determine if it contains an answer to the question. If it has produced a conflict, then the conjunction was proved by the current state. Thus an affirmative answer can be given. Furthermore, the condition can be relaxed to $c \vee e$, where e is the *exception* that created the conflict. In this way, an "any other?" question can be answered. In our case the relaxation may yield for a known entity a_1 in the domain:

$$\{appointment(x) \rightarrow x = a_1\}$$

Thus, the original constraint can be replaced with the relaxed one.
If, on the other hand, the reasoning task finished with producing a model (i.e. the set of requirements was satisfiable), the conjunction could not be proved and thus a negative answer **Done** has to be output to the user. In any case, after the task has been inspected, it can be discarded. Thus, once the system responds with a **Done** move, the current reasoning state is restored to the setting before the user question. In addition, the question is removed from /SHARED/QUD.

Queries containing more than one variable are handled in a similar way, by adding a conjunction when the clause is relaxed. The conjunction has to be implemented using an auxiliary predicate. For instance, the move (**Query** x, y $pred(x, y)$) would generate relaxations in the following form:

$$\forall x, y . pred(x, y) \rightarrow ex_1(x, y) \vee \ldots \vee ex_n(x, y)$$

Here, the predicates ex_i list the i-th tuple (x, y) matching $pred(x, y)$.
In addition to determining entities that match a user query, in order to generate the answers, a realistic system may also have to perform some additional presentation planning. This is to mean that in most domains, complex entities cannot be presented as such, but have to be described by their properties. In the calendar domain, for instance, the system response to the user question in U-1 should be "You have an appointment from 9 to 11" rather than "A1 is one of your appointments". For the kinds of domains we focus on in our dialogue management approach, we assume we can generate these kinds of descriptions from the domain information structure described in Section 8.4.

7.5 Providing Inferences

This section describes the introduction of inferences by the system. As mentioned above, the roles of the dialogue participants in our model are asymmetric in the sense that the user can state or drop requirements while the system may only react to these requirements.

In the following we address the explanation of inferences by the system. The system should be able to explain the inferences it has drawn on the basis of the user requirements and its background knowledge. In principle, this involves explaining the individual steps and rules that were used for generating the inference. Each inference has preconditions and possible alternatives which may need to be explained in order to present the overall inference step. Thus, in general this is a complex recursive problem.

In principle, the proof is always finite and can be translated into a series of statements describing it. However, the size of the proof makes this theoretical approach rather impracticable, since the user would most likely be overwhelmed quickly. Instead, the explanation should be an interactive process guided by the user's need for explanation. The user should be able to decide which parts of the proof are important or unclear to him.

The system's task is to communicate (relevant) inferences it has produced on the basis of its domain knowledge in conjunction with the requirements introduced by the user. This includes the communication of a detected inconsistency (i.e. a conflict), as a special case of inference. This is discussed separately in the next section.

In principle, the system communicates inferences by using an (Assert ϕ). In fact, that is the same assert move as used by the user to state requirements. However, a system inference move includes conditions in ϕ. Consider the following dialogue:

> U-1 > It's a two hour appointment starting at eight.
> (Assert $\{start(a_1, 8), duration(a_1, 2)\}$)
> S-2 > So, it ends at 10, right?
> (Assert $\{start(a_1, 8) \wedge duration(a_1, 2) \rightarrow end(a_1, 10)\}$)

The example is based on relatively strong assumptions concerning linguistic front-end. In particular, we assume that the "So,..." in S-2 will be generated from (and understood as) the preconditions in the Assert move. We discuss a prototypical approach to presenting inferences in the context of the prototype in Chapter 8.

The system side processing of its own Assert move is different from the processing of the respective user move. In the case of the user move, the system has to create respective reasoning tasks. In the case of a system move, the reasoning task does exist and is the basis of the move.

One challenge here is to ensure the coherence of the dialogue because inferences are produced by the reasoning engine in an asynchronous fashion. In particular, an inference may not be dependent on the requirement most recently introduced by the user. In addition, there may be different inferences to choose from in the dialogue. However, since the proof structures are available to dialogue manager, it may determine the relation of an inference with regard to the existing context.

The Gricean maxims of quality, quantity, relevance, and manner may serve as an orientation concerning the system output. The quality should not be a problem, since the system only provides information it has inferred. However, it should make sure that inferences and suggestions are not confused. Relevance may be addressed by relying on the proof structures to identify the relations of an inference given the user requirements. Manner can be assumed to be handled by the linguistic interface. The main difficulty is probably the question of quantity. There may be a lot of inferences and also the level of detail in these inferences may be too high to just output all.

7.5.1 Resolving Conflicts

The requirements stated by the user may be inconsistent with some information that may only be available or apparent to the system. In this case, it is the task of the system to explain the sources of the conflict in order to enable the user to revise some of the requirements. Thus, the user should be enabled to decide which of the involved requirements to modify or drop. One aspect already mentioned above is that the system inferences are to take place as soon as possible, even when only a partial description of a scenario or set of requirements is available. In this way, conflicts should be detected as early as possible in contrast to a *late detection* that takes place only when other possibly unrelated information has been added. Early conflict detection is desirable, because the conflict resolutions becomes more complicated once the dialogue has progressed to another topic. Furthermore, information provided later in the dialogue may be rendered irrelevant if earlier information is already conflicting. Thus, an early conflict detection may avoid unnecessary dialogue interactions that may occur in a late detection scenario. The presented approach enables an early detection, since any user requirement is processed as soon as it is introduced. In addition, the system strategy assigns a high priority to the handling of conflicts.

A conflict is a special case of an inference. It is handled by the general case.

For the purpose of resolution, a conflict can be interpreted as a proof of the negation of one of the user requirements that led to the conflict (i.e. the negation of one proposition is proved by contradiction from the remaining requirements). It is the decision of the system to choose which requirement to negate. As a baseline, the most recent requirement may be chosen.

For instance, in the calendar domain:

```
S > When does it start?
U > Ten.
S > When does it end?
U > Nine.
S > It starts at ten, so it cannot end at nine.
S > When does it end?
```

A form of *simple conflict resolution* takes place when the user states two different values of a property. In that case, the earlier requirement is negated, i.e. it is replaced. Thus, values can be corrected by the user by providing the correct value in an assertion. These situations can be recognised from the domain information structure and correspond to overwriting a field value in a VoiceXML form. Consider, for instance, the following example from the calendar domain:

```
S > When does it start?
U > Nine.
S > When does it end?
U > It starts at ten.
S > Ok.
```

7.5.2 Dealing with System Suggestions and Alternatives

Rather than providing inferences that necessarily follow from the requirements stated by the user, the system may also provide suggestions, or suggested assumptions, in various situations. This happens when the reasoning engine generates (potential or complete) alternative solutions which cannot be ruled out on the basis of the given information. The alternative solutions are then based on additional assumptions introduced by the reasoning process. Technically, the role of these additional assumptions is to saturate logical domain rules specifying a disjunctive postcondition. If a (partial) solution is to be discussed by the system, it has to communicate the underlying assumptions. It is thus important for the system to enable the user to distinguish between the necessary inferences based on user requirements and suggestions (for which alternatives exist). In particular, the user will typically be interested in knowing what alternatives to a proposed statement are possible.

In the following utterance, the system proposes to use a certain entity in the TRAINS domain:

```
S > (6) How about taking engine E1?
```

However, the system should not propose assumptions that the user explicitly wants to leave open, i.e. assumptions concerning decisions that the user does not want to take right now:

```
U > I want Italian or Asian food.
S > How about Italian food?
```

On the other hand, in the following situation, a resolution may make sense:

```
U > I want to eat Asian food today?
S > What kind of Asian food? How about Indonesian?
```

The two situations can be distinguished on the basis of the user requirements.

When the inference process comes up with more than one resulting scenario, the system has to decide how to best make use of this information. In a form-like dialogue, i.e. discussing properties of an object, it could just select the scenario it considers the best and present the respective information as a suggestion:

```
U > Add another appointment.
S > When does it start?
  > How about eight?
```

Alternatively, it could present choices:

```
S > When does it start?
  > How about eight, or any value between 10 and 17?
```

In that case, alternative restrictions are presented using "or".

7.5.3 Retracting Inferences

Concerning the retraction of information in the dialogue, two situations have to be distinguished: First, the retraction of requirements by the user (cf. Section 7.3) and second, the retraction of inferences by the system. The latter may happen when the user has retracted one or more requirements, because the system needs to retract the inferences that were based on these requirements. Of course, the system does not have to do anything if it has not communicated these inferences to the user yet.

```
U-1 > I am looking for a Bengali restaurant in the city centre.
S-2 > Ok, there is only one restaurant. The name is "Anondo".
U-3 > Well, then let's look somewhere else.
S-4 > Ok, then we have more options: ...
```

In utterance 1, the user asserts the requirement $location(r, Centre)$. In utterance 3, the user retracts this requirement and introduces the negated constraint $\neg location(r, Centre)$.

7.6 Using Hypothetical Reasoning

The ability to use a so-called *hypothetical mode* is one of the features that probably illustrate best the advantages of using a separate reasoning engine to manage sets of domain-level requirements. Using different sets of requirements is natural to the reasoning engine and is built-in at the core of the interactive protocol (cf. Section 5.4). It provides a solid basis for functionalities such as comparing scenarios or solutions.

As illustrated in the following example, the user can enter hypothetical mode and leave it either by committing to the hypothetically discussed topics (U-a), or by cancelling them (U-b):

> U > What if it started at ten?
>   (Enter $\{start(a_1, 10)\}$)
> S > It would end at 12.
>   (Assert $\{start(a_1, 10) \rightarrow end(a_1, 12)\}$)
> U-a > Ok, let's do that.
>   (Commit)
> U-b > Forget it.
>   (Cancel)

The usefulness of a hypothetical mode stems from two facts:

- First, hypothetical requirements can be disposed when no longer needed. In this case, the interaction returns to the previous state.
- In addition, requirements that are in conflict with the current situation can be discussed. In this case, the system performs a form of simple conflict resolution that prefers the hypothetically added requirements.

The main advantage of this functionality is to allow the user to hypothetically reason about certain assumptions and having a riskless possibility to return to the previous state of affairs. Alternatively, the user can accept the changes made in hypothetical mode as the current state. If a conflict arises by entering the hypothetical mode, the default behaviour of dropping the last requirements is overridden by preferring the last requirements instead and dropping any other requirements as needed. This makes sense since the user wants to reason about a new situation that is likely to be inconsistent with the current situation.

Dialogue moves concerning the hypothetical mode are integrated as follows:

- An (Enter ϕ) move is processed by creating the hypothetical mode structure in /CONTEXT/MODE. This also creates a backup of the current requirements in /SHARED/REQ. Then ϕ is added as a requirement in the same way an Assert move is processed. If a conflict arises, any conflicting previous requirements are retracted.
- A Commit move simply drops the hypothetical mode structure. The state of requirements discussed last remains active.
- A Cancel move, on the other hand, restores the requirements in the backup and then exits from the hypothetical mode.

On the linguistic level, the hypothetical mode is signalled by using the conditional tense. This information is represented in the CONTEXT portion of the information state. Technically, the dialogue move engine maintains a backup copy of the reasoning state within the information state. That backup can be re-activated or disposed of as required when leaving the hypothetical mode.

It may also make sense to allow the system to initiate a transition to the hypothetical mode, for instance, to present suggestions to the user. Using the hypothetical mode may indicate to the user that the system's utterance is a suggestion and not an inference.

Dealing with system suggestions can be implemented with the hypothetical mode moves. Essentially, a system utterance of the form "How about X?" initiates a transition to the hypothetical mode and asserting X.

The user may also initiate such a move in order to *explore* the different solutions. For instance, in a travel domain, the user may explore the solution space by choosing various combinations of destinations or means of transportation. In general, the user should be allowed to browse different solutions or potential solutions and evaluate them, e.g. by comparing their respective logical conditions. In principle, exploring different solutions can also be achieved using the mechanisms described for managing hypothetical reasoning. However, using the `Enter` and `Cancel` moves introduced there, while switching between scenarios is possible, only scenario can be inspected at any time. A direct comparison is not possible.

An example dialogue that illustrates discussing alternatives in the calendar domain is given in the following fragment:

```
U-1 > Add another two hour meeting today.
S-2 > Ok, when does it start?
U-3 > What are my options?
S-4 > You have an appointment from eleven till twelve.
    > Your appointment can be before that or after that.
```

In utterance `U-1`, the user adds another appointment to an existing schedule. Apart from the new appointment, there is an existing appointment, which results in two scenarios created by the reasoning engine. In `S-2`, the system asks for more information in order to resolve this ambiguity. Instead of answering directly, the user initiates a subdialogue explicitly asking about possible solutions. In utterance `S-4`, the system provides information on how the two scenarios were inferred by the reasoning engine.

7.7 Discussion

In this chapter we have described an extended approach to dialogue management based on the integration of an interactive reasoning engine. The approach has been developed to better address some of the restrictions that have become apparent during the implementation of the VoiceXML-based prototype (cf. Chapter 6). Instead of relying on the concepts that VoiceXML provides, and certain limitations that go along with them, this approach is better suited to taking advantage of the reasoning engine. It provides the essential functionalities offered by reasoner's interactive protocol to the level of dialogue management.

We have developed a generic system strategy based on the requirements-inferences approach. It explains the different priorities of dialogue functions. The modelling approach is based on the notion of dialogue moves and introduces new dialogue move types whose system-side processing has been described. Some of these moves, for instance assertions, map directly to operations based on the interactive protocol. Others are rather concerned with the management of the dialogue interaction, for instance, regarding the current focus of attention. The integration of the described dialogue move engine into a complete architecture and prototype system is the topic of the next chapter.

Concerning our dialogue management approach, a number of possible extensions and enhancements remain to be discussed.

For instance, solutions are treated rather differently than domain entities in the dialogue. The technical reason is obvious: solutions are not domain entities; if anything, they are meta-domain (or problem-solving) entities. However, in terms of interaction, it may make sense to treat them in a fashion that is more similar to domain entities. For instance, it may make sense to explicitly list solutions in the same way that domain entities can be enumerated (cf. Section 7.4). In particular, it may make sense for a user to specify certain attributes to refer to the "simplest" or the "cheapest" solution. In general it is non-trivial to explicitly refer to different scenarios using natural language utterances. One approach may be to derive descriptions from the contents of the scenarios, or from branches that have occurred in the respective reasoning process.

Other improvements may concern the way inferences are provided by the system. The current approach essentially delegates most of the work to a domain-specific handling. There are a number of general issues concerning the communication of system inferences: Firstly, determining which inferences are important: A conflict is apparently the most important inference, since it prevents further reasoning. However, concerning other inferences it may be necessary to develop domain-specific criteria. For instance, the inferred existence of certain types of events (e.g. travels or loading events in the TRAINS domain) may need to be reconfirmed with the user. In addition, determining the adequate level of detail of the presentation is difficult: On the one hand, the system should provide as much information as possible, on the other hand, it should avoid including "trivial" proof steps in order to keep the dialogue efficient. Finally, linearising the proof and determining the structure of the proof presentation can be achieved in different ways. For instance, the system should determine how the presentation can be chunked into smaller segments between which the user may assume the turn, ask questions to initiate subdialogues, or control other aspects of the presentation. The system should provide some functionality for summarising inferences by the system. This capability is related to the explanation functionality. Typically, not all inferences of a proof to be explained are equally relevant to the user. In fact, some inference steps may be rather trivial and thus need not be communicated.

The general system strategy has some degrees of freedom. For instance, more specific strategies may concern the proactiveness and the initiative of the system in the interaction with the user. In the next chapter, we describe how some of these issues are addressed in the revised prototype based on the information states approach.

8

Revised Prototype and System Architecture

8.1 Introduction

In this chapter, we describe the implementation of a revised prototype that has been developed on the basis of the information state-based dialogue management approach described in the last chapter. This chapter includes the description of a basic infrastructure for realising the integration between the reasoning engine CIDRE, the prototype dialogue manager, and components of the linguistic front-end.

The main goal of developing the prototype is to describe which domain-specific adaptations of the generic dialogue management approach may be required to build an end-to-end system. In particular, we describe a structure called domain information structure which encodes additional information about the domain. This information corresponds, to some extent, to the structure of a form in VoiceXML.

The main focus of the last chapter was to define the interface between the dialogue manager and CIDRE. Here, the interfaces between dialogue manager and the linguistic interface components are described on the basis of the approaches used in the prototype implementations.

In the following an end-to-end architecture is described as the proposed way to integrate the reasoning engine CIDRE. We illustrate the architecture by discussing the revised prototype in Section 8.6. This system features a calendar domain and a (text-based) interaction developed on the basis of the described dialogue management approach (cf. Chapter 7).

The proposed architecture comprises the following general components (cf. Figure 1.1): A linguistic front-end (bi-directionally) translates between natural language utterances and dialogue moves. It also controls the updates to the linguistic discourse context. The dialogue move engine maintains the information state, integrates observed moves into it and generates system dialogue moves. Based on the information state, reasoning tasks are initiated and monitored by the dialogue move engine. Results of the reasoning tasks are the basis of most system dialogue moves, for instance, system query and answer moves. The reasoning engine maintains a set of individual reasoning processes which are created by the dialogue move engine.

8.2 Architecture Considerations

This section discusses considerations that influenced the design of our proposed architecture. In many cases, the different components of an end-to-end dialogue system are implemented in different programming languages and frameworks. It is thus necessary to translate the communication between components with different native message formats. Many systems, such as DARPA Communicator [98], Verbmobil [99], or SmartKom [66], mainly use a text-based communication format. The advantages of text-based messaging include human readability and straightforward logging mechanisms. On the downside, generating and parsing text messages can constitute a significant overhead. Another difficulty is to communicate binary data, in particular, if the transmission is to be streamed. In this case the data is split into packets to be received individually rather than the whole block at once. A typical example of this kind of data is audio data, both on the receiving side (from an audio source, such as a microphone) and the sending side (to an audio sink, such as a speaker system). Here, it is also quite apparent that, in general, there is no pre-determined end point of the data stream. Even if there were, the transmission should not wait until it detects it, but instead send the available parts of the data as soon as possible in order to enable parallel processing. In order to stream binary data, the Communicator platform includes a special type of direct communication between pairs of components. A rather ad-hoc alternative approach to exchange larger amounts of data is to rely on a shared file system and to communicate only via file references in messages. The disadvantage of this approach is that important structural information is hidden from the communication system (and the logging facilities).

Other than text-based communication formats, projects have used middle wares like the Common Object Request Broker Architecture (CORBA) or Remote Method Invocation (RMI, Java). The advantage of this kind of communication is that it may require fewer modifications of existing code in order to enable its usage in distributed systems. In fact, these approaches attempt to hide the details of the communication from the user as much as possible. Systems like CORBA and RMI are also geared towards object-oriented components. I.e. they support methods calls and object references (rather than object copies) in an efficient way. One may argue that these kinds of communication formats may be more efficient than text-based messaging. On the other hand, these mechanisms incur specific problems of their own. For instance, they require more effort to provide logging support, essentially at this point a message-to-text conversion has to be implemented.

As the complexity of the system increases and the system testing gets increasingly extensive, the ability to conduct meaningful test cases in an efficient manner is a vital requirement to a developer involved in dialogue systems components implementation. One of the baseline approaches is to use a logging facility and to analyse and compare the logged data from different runs. In particular, developers may want to make sure that given the same sequence of inputs, the system produces the same processing results. Where available, the ability to automatically generate these input data for the system (batch testing) is an accepted and valued development tool.

8.3 A Basic Black-Board Infrastructure

The considerations outlined in the previous section have led to the following conclusions concerning the architecture of our prototype system. First, distribution is less important in our system because we focus on the integration between a dialogue manager and a reasoning engine. In addition, we have chosen the same programming language for these two main components, thus eliminating the need for translation between different representations. However, we are still convinced that it is useful to adhere to an architecture that distinguishes several processing modules with well-defined interfaces and an explicit communication mechanism that interconnects them. The main reason is that meaningful logging and, to some extent, simulation of processing components should be supported by the architecture.

The central components, including the dialogue move engine, the reasoning engine, and parts of the natural language front-end (i.e. parser and generator) have been implemented as separate modules in the Haskell programming language. At run time they operate as separate threads. These modules exchange messages via a shared black-board communication mechanism that allows each module to send and receive messages.

This approach has the following advantages: Firstly, a shared address space is used, i.e. the sharing of data structures is possible. In fact, we can achieve a tight integration of the processing components and efficient operation, since operating in the same process space implies that lazy evaluation can take place, i.e. messages are evaluated on demand and only as far as interpreted by the receiving module. This enables us, in the case of the parsing component, for instance, to send multiple parse hypotheses in a list structure without spending computational resources on hypotheses that are not evaluated by the dialogue manager. Thus, one is not required to limit the number of hypotheses to an arbitrary bound in advance, the number of hypotheses that can be evaluated is only determined by the parser's current input and grammar.

In addition, the messages are strictly and statically typed, i.e. type checking allows to avoid a variety of communication and conceptual errors that might otherwise be detectable only via extensive testing at run time. In fact, each processing component can be assigned a type signature based on the kind of input and output messages it processes.

All the processing components in our proposed architecture are derived from the main processing loop sketched in the code fragment in Figure 8.1. The basic processing cycle consists of receiving messages from a communication channel and processing these until a termination message is found. The processing includes sending messages to the shared black board. Being functions, each of the processing modules can be assigned a static type signature.

Require: $st \in$ ModuleState
1: **while** $msg \leftarrow$ receiveMessage **do**
2:    **if** isEOF msg **then**
3:       return
4:    **else**
5:       $(st, msgs) \leftarrow$ processMessage st
6:       sendMessages $msgs$
7:    **end if**
8: **end while**

Fig. 8.1. Code fragment to illustrate module processing.

The format of the messages is defined by a global abstract data type that distinguishes between the different structures required for each module. For instance, this data type defines the possible messages that can be sent to configure the natural language parser and the type of messages that will be sent by the parser when natural language input has been analysed. In a similar way, the communication between the dialogue manager and the reasoning engine is handled. The details of the messages used for the communication are described in Section 5.4.

Concerning the testing of individual processing components in the architecture, the approach taken is to use the black board infrastructure as a logging facility. In addition, processing modules can be simulated by sending pre-defined messages, for instance, messages recorded in a previous run of the system.

8.4 Frame-Based Surface Level

The interaction approach as described in terms of requirements and inferences in Chapter 7 is still rather abstract and general. For developing prototypes of our dialogue management approach in practical domains, we extend this approach with certain assumptions concerning the surface level of our dialogues.

In particular, we assume a form-based structure of certain classes of domain entities for the purpose of the dialogue interaction. However, we do not use a complete form-based control as in VoiceXML. Rather, the form-based structure refines the generic system strategy at a lower priority. For instance, the form-based structure is consulted by the dialogue manager when determining whether a specific system query should be raised.

Thus, we focus on application domains in which the main task of the user is to discuss entities of a certain form-structured class. In order to avoid confusion with the VoiceXML term, we refer to such a class as a *main class*, or *frame class*.

In a flight information domain, for instance, the common task of the user and the system is to filter a flights database according to the criteria defined by the user. Similarly, in the restaurant domain, the task is to choose a restaurant from a list, including choosing the cuisine from a hierarchy of options, and choosing the location in a similar manner. In the calendar domain, the user maintains a schedule of appointments, making sure that no conflicting appointments are created. Appointments can be created, modified, or dropped. Managing these entities concerns activities like updating the entities on new requirements, as well as browsing the current state of information. The main consistency rule is that certain appointments (i.e. those that require the same resources) cannot coincide.

Much of the dialogue in such frame-based domains is dedicated to discussing the properties of individual entities (of the frame class). For instance, when a new appointment is added, the starting and ending times have to be gathered, and possibly other choices (for instance, the location). Thus, respective requirements describe individual entities using some of their properties.

Requirements may also concern multiple entities. For instance, the user may require one appointment to take place after another, without specifying concrete time values. This requirement concerns multiple entities, not just the properties of one entity. Consequently, inferences and, in particular, conflicts may concern one entity or multiple ones. The dialogue has to distinguish between these cases.

One specific challenge concerns the natural language interaction: Speaking about different entities using natural language can be a difficult matter: Frame-based entities are often not referred to by an assigned name, but through descriptions that are to identify the entity on the basis of its properties or relations to other entities. For instance, in the calendar domain, related expressions may be "the other appointment", or "my appointment at 10". It is easier to discuss only one entity at a time (as in a VoiceXML form) than to discuss relations between such entities.

We argue that focusing our attention to frame-based dialogues does not limit our approach to a very small range of domains, and in particular, does not compromise our goal of integrating different domains. For instance, taking the calendar domain as a basis, one could imagine extending the application to be integrated with a restaurant domain, which could also be realised in a frame-based dialogue approach. The idea of combining these two domains would be that some types of appointments could be scheduled in a restaurant which in turn would be selected on the basis of certain related criteria. The calendar domain could also be integrated with a (public) transportation domain, or with a room management service in a business setting. In particular, in an application domain it is quite natural to define more than one frame class.

In order to implement a frame-based approach, we introduce a structure that captures the domain-specific information that is required. The *domain information structure* contains the information about those classes of objects in the domain that can be discussed with the user. It also contains information about how these objects are structured in terms of properties. One important aspect is that grammars are specified in order to handle the respective user input.

$$
\begin{bmatrix}
DomainInformationStructure \\
\text{CLASS} \quad\quad : \text{ ``appointment''} \\
\text{GRAMMARS} \quad : \ \langle \text{ ``appointment.srgs''} \ \rangle \\
\text{PROPERTIES} \quad : \ \left\langle
\begin{bmatrix}
\text{NAME} \quad\quad : \text{ ``start''} \\
\text{GRAMMARS} : \ <\ \ldots\ > \\
\text{NAME} : \text{ ``end''} \\
\ldots \quad\quad : \\
\text{NAME} : \text{ ``duration''} \\
\ldots \quad\quad :
\end{bmatrix}
\right\rangle
\end{bmatrix}
$$

Fig. 8.2. The Domain Information Structure.

8.5 Refining the System Strategy

System queries. In Chapter 7 we did not address the question how the system raises questions. Here, we provide one possible strategy based on the domain information structure (cf. Figure 8.2). When trying to determine the best question to be raised, the system proceeds in the following way: The properties that can be asked for by the system are specified in the domain information structure. In order to avoid unnecessary focus changes, which might lead to an incoherent dialogue, the currently discussed object is preferred if it has an open property value. If the current object does not require any Query moves, i.e. all its properties have values, the dialogue context provides other salient objects to switch the focus to. Concerning the natural language interface, the domain information structure assigns grammars to the individual properties to be asked for. These grammars serve as expectations concerning the input and also define its semantic interpretation.

Managing Focus. Being able to switch the focus from one entity to another is an important way to guide the dialogue. The basic approach we propose here is to provide a simple way to switch between the most salient entities at any time in the dialogue. In particular, the Focus_Other move can be used to switch between the two top-most entities (in the salience stack). If that is not sufficient, a query move can be used to explicitly bring other entities into focus. The dialogue engine's discourse context maintains the stack of salient objects in /CONTEXT/FOCUS. These may be referred to using a limited set of "special" linguistic expressions. This is illustrated in the following example:

U > Let's talk about the appointment object instead.
(Focus_Other)

Here, the expression "the other appointment" denotes the second-to-top object on the focus stack.

The referring expressions can be summarised as follows:

- "this object", "it": This refers to the object that is currently under discussion between the system and user. Also, the name of the class may be used, as in the example.
- "the other object", "the other one": As above, this expression refers to the object second object on the focus stack.

Maintaining the focus stack means that the system has to update the focus stack at certain points in the dialogue. In particular, when an explicit switch to the "other" object is performed, this object is pushed to the top of the stack and is now to be referred as "it". Conversely, the old top object is now the "other" one, so that the top two stack elements essentially swap positions. However, this update does not take place when the "other" object is only referred to, but not switched to, as in the following example:

U > This appointment starts at after the other appointment.

Another situation in which the focus stack has to be updated is when a new object is introduced. In this case, the new one will be the current object, while the former will be pushed back to be the "other" one.

8.6 The Revised Prototype

In this section we describe the specific components of the revised prototype that are necessary to extend our dialogue approach to a system to interact with. The prototype has been implemented using the black board architecture described in the previous section. The goal of this prototype is to define a minimal domain that exhibits substantial domain reasoning, to demonstrate the use of our dialogue management approach, and to provide a basis for future extensions to increase its usefulness. To this end, the prototype also contains a basic implementation of the linguistic interface components and knowledge sources.

8.6.1 Application Domain

The application domain of the prototype is essentially the calendar domain (cf. Section 4.4.8). It features appointment entities described by properties such as the starting and ending time. This is summarised in the following. We essentially deal with two types of entities, appointments and time values. Appointments are the frame class in this domain. Time values are auxiliary objects. The latter represent time instants and durations (periods). For our prototype, these time points have been further restricted to full hours.

The following basic domain rules are encoded in the domain theory (cf. Figure 8.3). Each appointment (apparently) has a starting and ending time with the starting time preceding the ending time. Two distinct appointments requiring the same resource may not overlap. In this case a resource may be a room, a person's presence etc. For a minimal domain, a dummy resource is stipulated that all appointments require.

$$appointment(a) \rightarrow timepoint(start(a)).$$
$$appointment(a) \rightarrow timepoint(end(a)).$$
$$appointment(a) \rightarrow timespan(duration(a)).$$

$$start(a, t_1) \wedge end(a, t_2) \rightarrow timepoint\_precedes(t_1, t_2).$$
$$start(a, t_1) \wedge duration(a, d) \wedge end(a, t_2) \rightarrow sum(t_1, d, t_2).$$

$$appointment(a_1) \wedge appointment(a_2) \rightarrow$$
$$a_1 = a_2 \vee precedes(a_1, a_2) \vee precedes(a_2, a_1).$$

Fig. 8.3. Excerpt from the domain theory of the prototype system.

The following extensions could enhance the minimal domain in a future system: Firstly, including complete date (calendar day) information and a finer grained time point resolutions (e.g. quarter hours or minutes) would make the calendar domain more useful. Another relatively straightforward extension – concerning the domain – would be to allow multiple users to schedule their appointments. Each person availability would then be modelled as a required resource of an appointment. More generally, resources such as rooms and (mobile or immobile) equipment could be modelled. Finally, it would be interesting to model different locations and, for instance, integrate with route planner. In addition, different types of appointments could be modelled, for instance, appointments that take place in a restaurant (with would be the basis for an integration with that domain). However, these domain extensions would imply much more substantial extensions on the linguistic level, and thus, for the time being, the scope of the prototype domain was restricted to an initial version.

Figure 8.4 illustrates the domain with a sample interaction.

```
S-1 > Welcome.
      (Greet)
U-2 > Add another appointment at nine.
      (Assert y {start(y, 9)})
S-3 > When does it end?
      (Query x {end(a₁, x)})
U-4 > Ten.
      (Assert {end(a₁, 10)})
S-5 > So the duration is one hour?
      (Assert {duration(a₁, 1)})
U-6 > Yes.
      (Assert {duration(a₁, 1)})
U-7 > Add another appointment.
      (Assert y {start(y, 9)})
S-8 > When does it start?
      (Query x {end(a₂, x)})
U-9 > It takes two hours.
      (Assert {duration(a₂, 2)})
S-10 > So, it starts at ten or after?
      (Assert {start(a₂, x) ∧ timepoint_precedes(10, x)})
```

Fig. 8.4. Example dialogue in the calendar domain.

In contrast to the restaurants domain in the previous prototype, here the user can discuss multiple entities of the *appointment* class. The dialogue aims to illustrate the use of **Assert** moves for stating requirements and providing inferences. In addition, **Query** moves are used by the system to obtain new requirements from the user.

8.6.2 Natural Language Understanding

This section discusses some implications on the natural language interface that are relevant to our implementation of the prototype and the approach taken to address these. The dialogue manager has an interface based on dialogue moves. Thus, the natural language understanding module provides its interpretation of an utterance in the form of dialogue moves. Similarly, the natural language generation facilities need to provide a way for producing natural language text from the moves generated by the dialogue manager.

In the prototype, NLU takes place in two phases. First, the utterance is parsed and a semantic representation is constructed. This representation is subsequently processed by a simple form of contextual interpretation. Essentially, this phase decides whether it refers to one of the currently focused entities or a new one. The semantic representation is adapted accordingly.

The implementation of the natural language input components of the prototype builds upon components developed for the VoiceXML framework described in Chapter 6. In particular, an SRGS parser with support for EcmaScript tags and grammars have been developed by the author in that context. For the Haskell-based black board architecture proposed in this chapter, a re-implementation of that parser was performed. The current prototype relies on typed input. However, the use of the Speech Recognition Grammar Specification (SRGS) [32] provides a way toward a speech-based interaction, for instance by implementing a modular spoken language front-end on the basis of a VoiceXML platform.

The basic forms of utterances that the SRGS grammar for the prototype deals with are analysed in the following way:

S > The appointment is from eleven till one in the afternoon
    $\{type : assert, start : 11, end : 13\}$

A large number of semantically similar utterances can be analysed with a relatively low number of simple grammatical rules. This semantic representation in the form of a record structure, which is the natural representation in the context of VoiceXML, is then translated using the discourse context into a dialogue move depending on the type of the utterance: (Assert x $\{starts(x, 11), end(x, 13)\}$). As mentioned above, this happens in the subsequent phase of contextual interpretation.

The discourse context interacts with the interpretation and generation of natural language. The primary task of the discourse context in the prototype system is to support the use of pronouns and implicit references to the entity that is currently discussed. For instance:

S > When do you want to start?
S > When does it start?
S > When does the appointment start?

In the first utterance there is only an implicit reference to the appointment entity we are talking about. In the second case, the entity is referred to using a pronoun. Finally, in the third utterance, the definite noun phrase "the appointment" is used, implying that there is exactly one such entity that we are talking about at the moment. The discourse context can also be updated by dialogue moves, for instance, in order to change the focus to another appointment.

8.6.3 Natural Language Output

One of the main features of our approach based on a domain reasoning engine is the availability of the information necessary to have the system explain its inferences in detail. The challenging question is how to use this information and present proofs in an adequate way. From the perspective of the dialogue manager we assume that the linguistic interface is able to realise the respective dialogue moves that are used to present inferences. In this section, we concentrate our description on the approach taken for the prototype.

In general, the task of proof presentation can be seen as a problem of natural language generation (NLG). NLG usually distinguishes between the following phases [100]: In the *document planning* phase, the content of the intended presentation is selected and its discourse structure, e.g. concerning argumentation, is determined. The subsequent phase of *micro-planning*, concerns the use of referring expressions, aggregation (conjunction, ellipsis), and lexical selection (for instance, with respect to the syntactic category). Finally, the *surface realisation* phase determines the linguistic details of the presentation, such as the use of active or passive mode, the correct word order, and the morphological realisation of the words. In our approach, content selection and discourse structure are mostly driven by the proof structure. Micro-planning and surface realisation are achieved by using presentation templates.

Compared to a more general problem of proof presentation, for instance in tutoring systems (e.g. [101]) our situation is simplified for two reasons. First, we assume that we do not have to discuss the axioms of the domain theory themselves, since these are assumed to be common sense. Second, the proof structures produced by the reasoner are inferences of ground atomic formulas. In addition, these have been generated from the same kind of antecedent proofs. This means that we do not have to consider certain complicated aspects, for instance quantified formulas.

Our approach to inference presentation constructs a presentation from a proof structure through the following stages: The first stage is proof simplification. Here, we omit proof steps assumed to be obvious. This concerns, for instance, simple arithmetic operations. In addition, subproofs may be reused in a proof at multiple locations. In this case, in all but the first instance, the subproof is replaced by a reference to its first occurrence. In the terminology of text generation, the proof simplification is part of content selection.

The next stage is conflict reformulation. Here, a detected conflict is viewed as a proof of a contradiction. Instead of just proving that some requirements are inconsistent, the conflict is reformulated in a way that the negation of one requirement follows from the other requirements. In terms of logic, a conflict can be stated in the following way:

$$a_1 \wedge \ldots \wedge a_n \to \bot$$

This is to say that the assumptions a_i taken together are unsatisfiable. For the presentation in natural language, one would rather use the logically equivalent form

$$(a_1 \wedge \ldots \wedge a_{i-1} \wedge a_{i+1} \ldots \wedge a_n) \to \neg a_i$$

for some chosen i, because it is considered more natural.

Finally, the wording stage determines how each step of the (simplified) proof is presented. The most important aspect here is to distinguish between user requirements, strict system inferences, and suggestions. Also, the generation of referring expressions, aggregation, and the lexical selection is performed in this stage.

For the prototype, these stages are additionally simplified, such that each proof is reduced to the user requirements it is based on. This means that intermediate proof steps are omitted because they can be assumed to be obvious to the user. In the sample calendar domain this seems practicable, since most of the inferences are rather basic and may be assumed to be performed by the user as long as the source requirements are recalled correctly. In addition, if the requirements the proof refers to are present in the latest moves, they will not be restated at all.

The following example presents a simple system inference:

```
U > Add a new appointment.
S > What is the start time?
U > Start time is eight, duration is two hours.
S > So, the ending time is 10?
```

The final surface realisation stage is realised as a very basic template-based generation procedure. Table 8.1 lists a small subset of the possible template choices. In the template column, variables (a, t) and brackets are meta symbols.

Type	Expression	Template	Comment
Atom	$start(a,t)$	"the start [of a] is t."	
	$start(a,t)$	"a cannot start at t."	Negated
	$duration(a,t)$	"a takes t."	
Term	a	"it"	Focused entity
	a	"the other one"	Other entity
	t	"at t o'clock"	Hour value
	t	"at t PM"	Hour value

Table 8.1. Templates for system output in the calendar domain.

The actual choice depends on a large variety of contextual conditions. Some of the most important ones in the prototype are:

- The type of the expression (e.g. an atom or term). Also, in the case of a term, it is important to determine whether the term is an extensional value.
- The grammatical function to realise, i.e. the category of the phrase (for instance, a sentence, noun, or adverb).
- The polarity and mode, i.e. is the statement negated or hypothetical.
- Additional knowledge from the inference process.

For instance, a generic predicate such as $leq/2$ should be expressed in different ways depending on the terms it applies to. In particular, when applied to time instants, as in the calendar domain, "being less than" should be realised using the expression "occurring before". In general, the utterances generated by the system should also correspond to what the system is able to understand.

8.7 Discussion

In this chapter, we have presented details about the implementation of system architecture and a revised prototype. The prototype within the new architecture aims to address various issues that have become apparent in the construction of the first, VoiceXML-based, prototype. Based on experiences with distributed architectures and the communication between different programming languages, the system is designed as a single-process system implemented in the functional programming language Haskell. We argue that this infrastructure allows for an efficient and expressive communication between the individual processing modules. This, in turn, enables us to concentrate on the essential aspect, namely, the communication between the dialogue manager and the reasoning engine CIDRE.

The architecture of the prototype consists of domain-independent and reusable processing modules, such as the dialogue move engine and the natural language parser. These are configured for the calendar application domain by domain-specific knowledge sources, such as the natural language grammars.

For the realisation of the dialogue manager, the concept of a frame-based domain structure has been introduced. This structure corresponds in parts to the usage of forms in VoiceXML. For instance, it helps to determine the order of system questions. However, the general system behaviour is governed by the generic strategy introduced in Chapter 7. The concept of a domain information structure is applicable to a variety of domains, and the details of the adaptations of the dialogue manager's strategy have been discussed.

With regard to the linguistic front-end components, approaches adequate for the purposes of a prototype system have been implemented. One aim of the prototype was to clarify the required interfaces to these components. Natural language understanding is based on a standardised formalism for recognition grammars. On the system output side, a basic template-based text generation procedure was implemented. To date, this is the module that requires the most domain-specific efforts.

Due to the prototypical nature of the system and due to the intentional restriction of the application domain, a number of generalisations and improvements may be addressed in a future prototype. Although for the current prototype, we restrict the natural language interpretation to relatively simple utterances, we note that the interface and the parsing mechanisms are ready for more complex utterances. In a grammar-based approach to complex utterances it is important to adhere to the principle of compositionality [69]. That is to say that the interpretations of more complex structures should be composed of the interpretations of simpler phrases. Whereas SRGS is somewhat restricted in expressive power concerning the surface level (for instance, it is essentially restricted to representing a finite-state network of words), EcmaScript tags provide a powerful way to construct the result structures once a parse has been produced. Compositionality is supported by SRGS grammars through the use of function objects. These can be used to represent λ expressions.

Another potential improvements concern the approach to proof presentation. The proof presentation aims at the following general properties: First, the level of detail should be adaptable to what is adequate for the user and the current situation. At the highest level of detail, the presentation should include every individual step of the process. Some steps in a proof may be considered trivial and should be ignored. Second, ideally, the presentation should be an interactive process in which the user can actively engage and "steer" the system's behaviour. The user might want to focus on certain aspects of the proof while ignoring others. This should be particularly useful, for instance, whenever the user is trying to "check" the assumptions that the proof is based on. Finally, the way the system produces output has implications on the understanding side because the user may be influenced by the way the system outputs its information and may use the same constructs. The capabilities for understanding user utterances should be (consistent with and) at least as powerful as the presentation capabilities, i.e. the system should be able to understand the constructions it used in its own presentations.

Another important challenge concerning the presentation of proofs is the presentation of utterances involving complex unresolved terms: These kinds of terms are generated in the reasoning process when Skolem functions are used in inferences. This process may be iterated and thus deep structures may be created. One way to approach the problem may be to introduce new entities by their relation to elements of the set of the entities under discussion. The relational information is available from the reasoning engine through the proof structures.

9

Conclusions and Future Directions

9.1 Summary

We have presented a novel architecture integrating domain-level reasoning with dialogue management in spoken language dialogue systems. The approach consists of an adaptation and extension of the conventional SLDS architecture in terms of a dialogue manager that takes advantage of a proposed interactive reasoning engine. The task of the reasoner is to process the domain-level information in a requirements-inferences approach. This means that with regard to the domain level the interaction between the user and system is characterised by the user asserting or retracting requirements and the system providing inferences. By relying on this approach, various dialogue activities can be realised, such as hypothetical reasoning and conflict resolution. In particular, both of these activities correspond to choosing adequate sets of user requirements. Using our approach the dialogue manager is able to present inferences and relate them to the particular user requirements the inference is based on. Thus, the basis for an explanation and interactive conflict resolution is provided. We argue that these functionalities should be an essential part of a flexible and user-friendly SLDS. Since the reasoning engine manages the domain-level information, it can also be considered an interface between the dialogue manager and application domain functionality.

When we talk about reasoning, we focus exclusively on the domain level. We do not reason, for instance, about problem-solving or discourse-level intentions. This allows us to characterise and limit the requirements that our domain models and inference mechanisms have to fulfil. In particular, we rely on first-order logic, which might not be sufficient for modelling intentions and beliefs.

Our reasoning engine CIDRE provides an interactive protocol to support an incremental and extensive communication with a dialogue manager. In this context, "incremental" means that the reasoner and the dialogue manager are able to promptly notify each other in case of important events. In the case of the dialogue manager, such an event may be the assertion of a new requirement that is to be taken into account by the reasoner. The reasoner, on the other hand, may communicate important inferences, such as a detected conflict in the user requirements. By "extensive" we mean that the engine communicates detailed information about its reasoning and that it can be controlled by the dialogue manager in many aspects. The information communicated by the reasoner includes any solutions found or conflicts detected, the inferences it has generated, their proof structures, and the hypotheses it is currently operating on. The proof structures, in particular, are important for the dialogue manager to relate an inference to the user requirements it is based on. The most important functions for the dialogue manager to control the reasoner concern the creation and management of individual reasoning tasks.

Internally, the reasoning engine implements a tableaux-based model generation procedure. This method has certain properties that make it attractive for use in an SLDS context. Firstly, the reasoning applied is in principle relatively natural and intuitive. By this we mean that most of the reasoning is performed by forward inference on simple facts, i.e. ground atoms. They are used as preconditions in rules that generate possible alternative postconditions, such that one of the alternatives has to hold in the solution. In particular, many of the rules are Horn clauses, there is only one postcondition to consider. The possibility to rely on a small set of inference mechanisms is desirable with respect to the explanation of the inferences in dialogue.

The actual rules that are used to generate new facts from known facts are determined by the domain modelling. For a given application domain, we refer to this rule base as the domain theory. The rules that constitute the domain theory are assumed to be common sense. This implies that the rules are not negotiated with the user, only the facts derived on their basis. In the application domains we have focused on the rules fall in two general categories. The first category of rules describes the static structure of the domain and the entities. The other category describes temporal processes and dynamic properties of these entities. Such a temporal process is, for instance, a move operation in a logistics domain. In order to be able to model these processes in a coherent way, a formalisation based on fluents can be applied in different domains. Both the fluent theory and the static modelling provide the basis for an integration of application domains. In our approach, the integration is based on the use of shared concepts, i.e. common classes of entities. For instance, in a logistics domain and in a calendar domain time values are represented by the same class of entities. One of the results of this work is that it is possible to model different domains in a logic-based fashion. It has also been shown that this modelling lends itself well to generalisation, and thus is reusable for other domains.

We have demonstrated that the strength of our architecture lies in its conceptual simplicity, since it is based on the well-known formalisms and techniques and extending these where necessary. Our common interactive domain-level reasoning engine, CIDRE, aims to be assistive rather than directive in the sense that it does not claim to autonomously solve all tasks possible in the application domain. Instead, it contributes as much as possible and assumes that its contribution furthers the problem solving process in the sense that the user can build on it. This, in turn, may yield new opportunities for the reasoner in the next iteration of the interactive process.

In the remainder of this chapter, we discuss different aspects of our work and point towards directions for further research.

9.1.1 Dialogue Management Approach

Our goal was to provide a dialogue paradigm based on interactive reasoning that aims to leverage both the user's and the system's complementary strengths. In particular, the goal was to enable the user to provide information in a flexible way and to take advantage of the system's assistance with respect to the given information.

In this context, flexibility refers to the following qualities of the interaction: First, the user should be enabled to provide information on a level of abstraction or detail that is adequate. In particular, the user should be allowed to make use of logical connectives, such as negation. In addition, using hypothetical reasoning seems an essential part in flexible interactions, since it provides the user the ability to evaluate and decide between alternative scenarios. Finally, the flexibility implies a mixed-initiative style of interaction, such that the user can determine the order and the amount of information to provide or let the system provide assistance. In this context, assistance concerns the following functionalities. First, the system shall be able to answer queries based on the information provided by the user. In addition, the system shall cope with overconstrained situations that arise from logically inconsistent information. Similarly, underconstrained situations arising from incomplete information shall be handled. Finally, the system shall provide its assistance in a transparent way, i.e. relating its actions to the information provided by the user.

We argue that our proposed dialogue management approach accomplishes these goals to a large extent. The interaction paradigm proposed is based on user requirements and system inferences. User requirements allow a flexible and expressive way to state information. Relying on First-Order Logic, both basic and compound logical formulas may be used in requirements. The expressiveness of FOL seems adequate for representing wide ranges of user input [68, 69]. Basic formulas may be disambiguated in cooperation with the system in the sense that abstract predicates may be resolved into more detailed alternatives. For instance, the desired cuisine in the restaurant domain may be described by a high level category. Compound formulas can be composed out of basic expressions. In particular, negation can be used to restrict the solution space.

Within the framework of requirements and inferences the system provides its assistance with the help of our proposed interactive reasoning engine. In particular, the information state-based approach described in Chapter 7 explains how a proposed set of dialogue moves can be realised on the basis of the reasoner's interactive protocol. These dialogue moves can be used to provide the system functionality in a transparent way. In particular, the system can answer user queries by requesting proofs from the reasoning engine. This is similar to the handling of goal clauses in Prolog. In addition, it is able deal with overconstrained situations. To this end, the reasoner can detect which ones of the user requirements are conflicting. Thus, the basis for an explanation and conflict resolution is provided. The system can also deal with underconstrained situations, i.e. situations in which multiple solutions are possible. In this case the reasoner builds alternative solutions and also provides information regarding which alternative hypotheses the solutions are based on. The solutions can then be browsed or refined. Our generic system strategy based on the dialogue moves provides a basic mixed-initiative interaction which can be adapted for specific application domains.

9.1.2 Interactive Reasoning

Our goal was to provide a reasoning engine that realises the functions necessary to provide the basis for the proposed requirements-inferences dialogue management approach. In particular, in the SLDS architecture the interaction between the dialogue manager and the reasoning component should play a central role.

We argue that our proposed reasoning engine CIDRE provides one way to achieve the required functionality. In particular, based on its interactive protocol, the reasoning engine supports a mixed-initiative style of interaction. One specific issue in this context is incrementality. This means that the reasoner should support both incremental output and input. Regarding the reasoner output this is realised by generating a stream of messages consisting of inferences and additional information. Therefore, any client to this reasoning stream can be notified by important events, such as a detected conflict or branch, with very little delay. On the input side, a component controlling the reasoner can create new reasoning tasks that can be incrementally defined on the basis of existing tasks. Thus, for situations like added user requirements the system-side processing can be more efficient. In addition, our proposed interactive protocol also provides extensive access to the reasoner's processing. One of the most important aspects here is that each of the inferences generated are described by a proof structure. This proof structure provides the information required to be useful for the dialogue manager. In particular, the inference can be related to the user requirements it is based on. In addition, the proof structure provides the basis for a more detailed explanation facility.

Technically, the interactive use of the reasoner is implemented by the design of the reasoner as an iterative function. This function refines a reasoning state structure until it reaches a fixed point. Being implemented on the basis of this function, reasoning processes are interruptible and can be modified in

many ways. The iterative function realises a tableaux-based model generation method which seems adequate for the sorts of reasoning tasks that are
required in an SLDS context. In particular, the resulting logical models can
be interpreted and accessed efficiently. Thus, they can be the basis for system
proposals in the dialogue. In addition, the inference method relies on a limited set of inference rules. This supports, we argue, the system's possibility to
provide explanations of inferences.

The reasoning is based on the domain modelling approach presented in
Chapter 4. One goal of the modelling approach was to provide the basis for
an integration of different application domains. To this end, a number of
domains and generalised base concepts has been modelled in our logic-based
approach. One of the aims of the modelling is the (abstraction and) reuse
of the concepts modelled. This happens on two levels. On the first level of
integration, different domains agree to use the same base classes of entities,
in much the same way as different software programs agree on base types
and data structures. On the second level, a common methodology for certain
common tasks is applied. In our case, this refers to the application of fluents.
Fluents are widely applicable among domains, representing value properties
of objects that may be changed by events over time. Examples include the
location of a traveller or logistics object, or the status of an abstract object,
such as an appointment.

A building block in the integration of application functionality is the ability of the logic-based approach to define translations between different forms
of representation (via consistency rules). Thus, the applications may process
their business logic in the representation that is natural for them, while the
important aspects and inferences that may influence other applications are
translated into a shared representation which can be translated into representations that other applications can work with. Conversely, from the view point
of an application, important information obtained from other applications can
be translated into the application's internal format. In essence, the translation
(that is provided by our reasoning engine applying consistency rules for the
domains in question) is a form of synchronisation between different representations.

Another aspect related to the integration of domains is the access to external data sources. However, not all application functionality can be directly
interpreted as cleanly as rows in database tables. Application knowledge may
include heuristics that are difficult to express in logics. Sometimes these heuristics would directly influence the reasoning strategy, rather than provide inferences.

9.1.3 Architecture Prototypes

Two prototype systems based on our proposed architecture have been developed. In terms of the general architecture, both prototypes present possible

ways to enable an interaction between the dialogue manager and the reasoning engine CIDRE. Both prototypes operate on the basis of the interactive protocol provided by the reasoner. However, architectures differ in significant ways. Our VoiceXML prototype is realised within the framework of a specific VoiceXML-based platform. This platform enables the integration of (Java-based) software in order to make it directly accessible from script code within the VoiceXML dialogue. Using this approach, it is possible to extend existing VoiceXML applications with the reasoning capabilities provided by the engine. The revised prototype, on the other hand, features a more modular and more efficient infrastructure. The infrastructure supports the clean separation of system processing into modules which communicate by sending typed messages via a black-board mechanism. This mechanism enables a more efficient processing since it operates within one address space and can take advantage of features like lazy evaluation. In this architecture, one central component is the dialogue manager which implements the information state-based approach described in Chapter 7.

In terms of the functionality within their respective application domains, two different strategies have been pursued. The VoiceXML prototype attempts to add CIDRE-based functionality while remaining within the general VoiceXML framework. In particular, this provides the basis for enhancing existing VoiceXML applications. The prototype's application domain is a restaurant selection according to user-provided criteria which are represented in a form. The integration of CIDRE provides added flexibility for user input and a way to separate the domain-level processing from the dialogue, i.e. VoiceXML code. However, due to limitations related to the form-based representation, this approach cannot take advantage of the full set of capabilities of the reasoner's interactive protocol.

Our revised prototype, on the other hand, builds on the dialogue management approach based on information states and dialogue moves. Thus, it is able to accommodate a greater variety of dialogue behaviours, such has hypothetical reasoning. The prototype's application domain is calendar management, which provides a larger range of possible user requirements and system inferences. In order to support the coherence of the dialogue, the prototype relies on a form structure in the style of VoiceXML for describing certain classes of objects, in this case appointments.

Our logic-based architecture aims to make hypothetical reasoning easier to integrate by providing it as a built-in feature of the approach. By "built-in" we mean that all the reasoning the reasoning engine performs is hypothetical.

9.2 Future Work

The work presented in this book raises a number of interesting questions and points to directions for further research. These will be discussed in the following.

Linguistic interface. A more complete integration into an end-to-end dialogue system will involve substantial efforts regarding natural language understanding and natural language generation. However, such an integration may also provide the basis for conducting user tests to evaluate the effectiveness of our dialogue management approach. On the understanding side, the goal should be to take advantage of the flexibility that the logic-based approach offers, for instance, the use of logical connectives and operators. Domain-level reasoning may also contribute to disambiguating natural language input, for instance, concerning anaphora resolution [68].

On the output side, the integration should aim at being able to express (or approximate) in natural language every inference that the reasoner generates. Conflicts, in particular, are inferences that need to be communicated and explained on the basis of how they have been inferred. The explanation may be an interactive process. This should allow the user to direct the focus to the relevant parts of the proof structure. The major challenge concerns the fact that even though the individual steps in a proof are quite comprehensible, the overall proof structure can become immensely complex and thus difficult to understand. One factor of complexity may be the presence complex nested terms created during the reasoning process. This is discussed in more detail in the next paragraph.

Generation facility. In our architecture the goal of achieving a transparent system behaviour relies on the proof structures generated by the reasoning engine. These consist of relatively intuitive steps which, in principle, can serve as a basis for an explanation functionality. However, apart from the relatively limited domains we have investigated in our prototype systems, realising an explanation functionality is a challenging matter. In particular, the proof structures may grow extensively and may thus become arbitrarily complex. In fact, the reasoner cannot guarantee to generate only the smallest proofs possible. The size of the proofs also depends on how the domain is modelled. Our proposed approach aims at a domain modelling which is close to common sense, i.e. a modelling that realises concepts that can be assumed to be to some extent intuitive. However, sometimes the modelling makes use of auxiliary technical concepts which may provide a more concise modelling, but may be less intuitive for non-experts.

In addition, a problem of size can also occur on a different level. Not only can proof structures consist of a large number of individual proof steps, but also the individual inferred facts can be complex in the sense that they contain deeply nested term structures. These are usually created during the reasoning process by an iterated application of Skolem functions. For instance, in the TRAINS domain proofs may contain terms roughly corresponding to expressions like "One hour after the oranges arrive to be processed". This term may have the following structure: $sum(1, end(L(loc(E_1), start(N(juice(or_1), 0)))))$. Here, or_1 refers to some package of oranges, $juice(or_1)$ is the fluent that represents the state of the oranges (solid, or produced into juice), $N(..., 0)$ stands for the first event affecting this fluent, etc. While these kinds of terms are necessary in many situations, they are apparently difficult for humans to understand. A possible way to address this issue may be an interactive explanation approach in the sense that more complex terms are introduced in a step-wise manner which relates them to the terms already discussed.

Other modalities. Furthermore, an integration beyond the domain of spoken language dialogue systems seems promising: The reasoning engine might also be useful as a back-end of a graphical or multimodal interface. In many situations aspects of the reasoning state or inferences may be adequately represented by graphical displays (such as maps or time lines). In these cases, using displays may be more efficient than speech-only communication. Complex transportation plans in the TRAINS domain may be one example. In addition, the applicability of the reasoning engine in traditional graphical user interfaces may be investigated. For instance, domain-level reasoning may provide explanations to the user why certain widgets are disabled. Such information may help the user to understand the logic behind an application.

User preferences. Another extension, useful for both speech-based or graphical interaction, is the management of user preferences and priorities. In particular, a user should be allowed to state the level of importance of requirements, or adapt these accordingly in the course of the interaction. For instance, priorities would make sense in situations where conflict resolution is to take place. User preferences might specify which requirements to drop and which ones to keep. However, the situation is probably more complex than just adding a weight to each requirement, because it is the resulting overall scenario that matters to the user. Thus, even if the resulting scenario is compatible with all relevant requirements, it may turn out to imply further conditions which are not intended by the user.

Back-end integration. Effort may also be invested in achieving more integration at the back-end, i.e. implementing interfaces to more realistic domain data sources. For instance, a route planner may be integrated as a special type of inference producer. It may infer the details of a certain route, given its origin and destination. The important aspect is to retain the semantics of an inference, i.e. the information produced by the route planner should be recorded as a logical consequence of the relevant input facts or assumptions. Also, concerning the reasoner's processing, domain-specific heuristics for guiding the finite model generation procedure may be implemented. This would potentially enable the reasoning engine to act more autonomously. In the logistics domain, for instance, there may be strategies for planning transportation events that work well for most situations.

Multi-party collaboration. Finally, techniques to enable groups of people to collaborate seem to become increasingly important. Our work may be useful in this area in two contexts: First, in a remote collaboration the reasoning engine may provide a shared back-end. Such a back-end may be accessed by separate instances of a dialogue manager, one for each collaborator. Thus, for instance in an appointment scheduling domain, groups of people may effectively and efficiently contribute on the basis of their respective requirements. As discussed, the reasoning engine in particular provides the way for relatively expressive formulations of the requirements. For instance, uninstantiated terms and ordering constraints may be used to express spaces of options. The other context is the potential use of the system in a direct multiparty situation. In this case, the dialogue management approach may need to be adapted in order to represent the information states of the collaborators.

10

Additions

10.1 Description Logics

Description logic (DL) is a restricted form of first-order logic that has been popular for representing knowledge in various application areas, such as engineering or bioinformatics. In particular, description logic is connected to the term "semantic web" which is an effort to establish standard knowledge representation formalisms and algorithms (e.g. the Web Ontology Language, OWL, [102]) to enable a distributed and automated processing of information published on the web.

DL can be applied to encode terminological and structural information, i.e. the definition of classes of entities and the relations among them. For instance, the concept "mother" may be defined as a female person who has at least one child. Here, the "female" and "person" are classes and "child" is a binary relation. Within a specific application domain, such knowledge is often referred to as an *ontology* of that domain. In addition to terminological reasoning (i.e. reasoning about classes of entities), DL systems can be used to represent knowledge about individual entities.

One of the main goals of the DL approach is to provide as much expressive power as possible while maintaining as much of computational tractability as possible. In particular, while first-order logic is undecidable in the general case, in DL specific restrictions are imposed to guarantee decidability.

So, one may wonder if DL might be used as knowledge representation formalism in this work. The advantages of doing so would include the adoption of a standard methodology including a large body of research and existing tools and technologies. However, the main reason why it is not possible to use description logic as the only knowledge representation mechanism is the fact that the expressive power of DL is too limited in specific cases which have been found to be of interest for our work. In particular, in standard DL languages it is not possible to express relationships between property values of different objects. Consider, for instance, the notion of overlapping appointments in a calendar domain. The definition of this concept would require to relate the start and ending times of different appointment objects. Rather than restricting the expressive power in our work we have chosen to restrict what may be assumed to be computed automatically by the system. In fact, we view the construction of a shared solution as an interactive process between a human user and the automated system.

10.2 SHRDLU

SHRDLU is a computer programme written by Terry A. Winograd in the late 1960's. It enables typed natural language dialogues in a simulated blocks world domain. The user may issue commands such as "find a block which is taller than the one you are holding and put it into the box" and the system would respond with a clarification subdialogue concerning the reference introduced by the anaphoric expression "it". Thus, the system is capable of impressive interactions, but it has also a couple of important limitations:

- The system is limited to the blocks world domain.
- The internal representation of the formal rules governing the domain are unclear.
- The system seems to be somewhat brittle and complexely intertwined. Concerning the first item, although the system accepts astonishingly complex utterances, it apparently does not accept others which are some small variations. There are also sentences which produce an internal error when processed.

In contrast to our work, a large part of the effort seems to be devoted to the understanding of natural language, taking into account syntactic, semantic, and contextual information. Our work does not address these issues at this level. Rather, our work aims to investigate the interface between a domain-independent interactive reasoning component and an abstract dialogue manager.

10.3 Dialogue Systems As Proof Editors

Ranta and Cooper (2001) address the issue of implementing a dialogue system on the basis of a specific type of reasoning, namely type inference. The domain is information-seeking dialogue, where the system takes the role of an assistant that tries to help a human user reach a certain information goal, such as knowing the details of a possible trip itinerary. In the trip planning application, the user would supply general constraints like departure and destination locations and temporal restrictions, for instance. The machine's goal, on the other hand, is to provide the details as obtained from a domain-specific database. To this end, the machine has to determine which pieces of information it needs to be able to access the a domain database.

Ranta and Cooper show that type inference may be a valuable tool for implementing the system's operations. A system that allows a user to construct an object of a given type through a series of refinement or undo operations is called a *proof editor*. For instance, in the trip planning domain, the task of booking a trip has the following type signature:

```
book :
City -> City -> (v:Vehicle) -> Class v -> Date -> Booking
```

This formula represents the domain-specific knowledge that an object of type `Booking` can be obtained from a sequence of parameter objects, namely two instances of type `City`, one of type `Vehicle`, hereafter referred to as `v` in the dependent type of `Class`, plus one instance of type `Date`. Although the arrow symbol may seem to indicate otherwise, the order of these parameter objects is not necessarily relevant.

Comparing the idea of using a proof editor as an underlying inference strategy with our own approach, one can note the following: Using a proof editor as demonstrated by Ranta and Cooper provides an elegant means to achieve essential dialogue system behaviours, such as incremental construction of a solution, including the contribution of system inferences, resolving resulting conflicting information, and the ability to handle different forms of initiative. For instance, in the trip planning domain, when the user asserts `Business` as the value of the `Class` object, the system infers that the `Vehicle` object is `Flight`, since `Business` and `Economy` are `Flight` class categories, and not, for instance, `Train` classes. Type checking is able to represent these kinds of relations and perform respective inferences. If the user had previously chosen `Train`, a conflict arises. This situation is communicated to the user, who may then decide on undo operations to resolve it. What remains unclear is the expressivity of type encodings. Ranta and Cooper speak of "simple dialogues". In our work, in contrast, we strive for a logic-based knowledge representation that is able to represent complex domain-specific rules and facts, for instance, relating multiple structured event objects to each other.

10.4 Closed World Assumption

The *closed world assumption* is an assumption that is used in logic programming to state that the facts explicitly represented are exactly the facts that hold in a given scenario. This is to say that ...

The closed world assumption applies that a domain modelled is modelled exhaustively. For instance, in a trip planning domain, all kinds of transportation means (such as flight, train, car, etc.) are modelled.

A practical implication of the closed world assumption is that ...

The closed world assumption is related to the concept of *negation as failure* which signifies that a goal of a negative literal can be considered proven if its negation (a positive literal) cannot be proven as a goal.

References

1. Guido, R.C., Deng, L., Makino, S.: Special Section on Emergent Systems, Algorithms, and Architectures for Speech-Based Human-Machine Interaction. IEEE Transactions on Computers **56**(9) (2007)
2. Jurafsky, D., Martin, J.H.: Speech and Language Processing: An Introduction to Natural Language Processing, Speech Recognition, and Computational Linguistics. Prentice Hall (2000)
3. McTear, M.F.: Spoken Dialogue Technology – Toward the Conversational User Interface. Springer (2004)
4. Rabiner, L.R.: A Tutorial on Hidden Markov Models and Selected Applications in Speech Recognition. Proceedings of IEEE **77**(2) (1989) 257–285
5. Allen, J.F.: Natural Language Understanding. Benjamin/Cummings (1998)
6. Jelinek, F., Lafferty, J., Magerman, D., Ratnaparkhi, A., Roukos, S.: Decision Tree Parsing Using a Hidden Derivation Model. In: Proceedings of ARPA Workshop on Human Language Technology, Plainsboro, NJ (1994) 260–265
7. Jelinek, F., Lafferty, J., Mercer, R.: Basic Methods of Probabilistic Context Free Grammars. Speech Recognition and Understanding. Recent Advances **75** (1992) 345–360
8. Minker, W., Bennacef, S., Gauvain, J.: A Stochastic Case Frame Approach for Natural Language Understanding. In: Proceedings of International Conference of Speech and Language Processing, ICSLP, Philadelphia, PA (1996) 1013–1016
9. Minker, W., Gavaldà, M., Waibel, A.: Stochastically-based Semantic Analysis for Machine Translation. Computer Speech and Language **13**(2) (1999) 177–194
10. Bühler, D., Vignier, S., Heisterkamp, P., Minker, W.: Safety and Operating Issues for Mobile Human-Machine Interfaces. In: International Conference on Intelligent User Interfaces (IUI), Miami (USA) (2003) 227–229
11. Ferguson, G., Allen, J.F.: TRIPS: An Integrated Intelligent Problem-Solving Assistant. In: Proceedings of 15th National Conference on Artificial Intelligence (AAAI), Madison, WI (1998) 567–572
12. Qu, Y., Green, N.: A Constraint-Based Approach for Cooperative Information-Seeking Dialogues. In: Proceedings of International Natural Language Generation Conference, INLG02, New York, NY (2002)

13. Qu, Y., Beale, S.: A Constraint-Based Model for Cooperative Response Generation in Information Dialogues. In: Proceedings of AAAI-99. (1999)

14. Allen, J.F., Byron, D., Dzikovska, M., Ferguson, G.M., Galescu, L., Stent, A.: An Architecture for a Generic Dialogue Shell. NLENG: Natural Language Engineering, special issue on Best Practices in Spoken Language Dialogue Systems Engineering **6(3)** (2000) 1–16

15. Cooper, R., Larsson, S.: Dialogue Moves and Information States. In: Proceedings of 3rd International Workshop on Computational Semantics (IWCS-3), Tilburg (Netherlands) (1999) 398–400

16. Austin, J.L.: How to do Things with Words. Oxford University Press (1962)

17. Searle, J.R.: Speech Acts: An Essay in the Philosophy of Language. Cambridge University Press (1969)

18. Grice, H.P.: Logic and Conversation. In Cole, P., Morgan, J.L., eds.: Syntax and Semantics. Volume 3: Speech Acts. Academic Press, New York (1975) 41–58

19. Cohen, P.R., Perrault, C.R.: Elements of Plan-based Theory of Speech Acts. Cognitive Science **3** (1979) 177–212

20. Perrault, C.R., Allen, J.F.: A Plan-Based Analysis of Indirect Speech Acts. American Journal of Computational Linguistics **6**(3-4) (1980)

21. Grosz, B.J., Sidner, C.L.: Attention, Intentions, and the Structure of Discourse. Computational Linguistics **12**(3) (1986) 175–204

22. Pollack, M.E.: Plans as Complex Mental Attitudes. In Cohen, P.R., Morgan, J., Pollack, M.E., eds.: Intentions in Communication, MIT Press (1990) 77–103

23. Grosz, B.J.: Collaborative Plans for Group Activities. In: Proceedings of 13th International Joint Conference on Artificial Intelligence (IJCAI), Chambery (France) (1993) 367–373

24. Grosz, B.J., Kraus, S.: Collaborative Plans for Complex Group Action. Artificial Intelligence **86**(2) (1996) 269–357

25. Lambert, L., Carberry, S.: A Tripartite Plan-Based Model of Dialogue. In: Proceedings of 29th Annual Meeting of the Association for Computational Linguistics, Berkeley, CA (1991) 47–54

26. Chu-Carroll, J., Carberry, S.: A Plan-Based Model for Response Generation in Collaborative Task-Oriented Dialogues. In: Proceedings of 12th National Conference on Artificial Intelligence. Volume 1., Seattle, WA (1994) 799–805

27. Chu-Carroll, J., Carberry, S.: Conflict Resolution in Collaborative Planning Dialogues. International Journal of Human-Computer Studies **53**(6) (2000) 969–1015

28. Ramshaw, L.A.: A Three-Level Model for Plan Exploration. In: Proceedings of 29th Annual Meeting of the Association for Computational Linguistics (ACL). (1991) 39–46

29. Sidner, C.L.: An Artificial Discourse Language for Collaborative Negotiation. In Hayes-Roth, B., Korf, R., eds.: Proceedings of 12th National Conference on Artificial Intelligence, Menlo Park, CA, AAAI Press (1994) 814–819

30. Ericsson, S., Lewin, I., Rupp, C., Cooper, R.: Dialogue Moves in Negotiative Dialogues. Technical Report SIRIDUS Project Deliverable (2000)

31. Larsson, S.: Issue-based Dialogue Management. PhD thesis, Department of Linguistics, Göteborg University, Göteborg (Sweden) (2002)

32. Oshry, M., Auburn, R., Baggia, P., Bodell, M., Burke, D., Burnett, D.C., Candell, E., Carter, J., McGlashan, S., Lee, A., Porter, B., Rehor, K.: Voice Ex-

tensible Markup Language (VoiceXML) Version 2.1 W3C Recommendation. W3C – Voice Browser Working Group, www.w3.org/TR/voicexml21. (2007)

33. SALT-Forum: Speech Application Language Tags (SALT) – 1.0 Specification. http://www.saltforum.org-/saltforum/downloads/SALT1.0.pdf (2002)

34. Aust, H.: Sprachverstehen und Dialogmodellierung in natürlichsprachlichen Informationssystemen. PhD thesis, Mathematisch-Naturwissenschaftliche Fakultät, Rheinisch-Westfälische Technische Hochschule, Aachen (Germany) (1998)

35. Hennecke, M.E., Hanrieder, G.: Easy Configuration of Natural Language Understanding Systems. In: Proceedings of the International Workshop Voice Operated Telecom Services, COST 249, Gent (Belgium) (2000) 87–90

36. Matheson, C., Poesio, M., Traum, D.: Modelling Grounding and Dialogue Obligations Using Update Rules. In: Proceedings of 1st Annual Meeting of the North American Association for Computational Linguistics (NAACL2000). (2000)

37. Lappin, S., ed.: Interrogatives: Questions, Facts and Dialogue. In: The Handbook of Contemporary Semantic Theory. Blackwell Publishers (1996) 385–422

38. Carpenter, B.: The Logic of Typed Feature Structures. Cambridge Tracts in Theoretical Computer Science 32. Cambridge University Press (1992)

39. Larsson, S., Traum, D.: Information State and Dialogue Management in the TRINDI Dialogue Move Engine Toolkit. Natural Language Engineering, Special Issue on Best Practice in Spoken Language Dialogue Systems Engineering 6 (2000) 323–340

40. Bos, J., Klein, E., Lemon, O., Oka, T.: Dipper: Description and Formalisation of an Information-State Update Dialogue System Architecture. In: Proceedings of 4th SIGdial Workshop on Discourse and Dialogue, Sapporo (Japan) (2003) 115–124

41. Allen, J.F.: Dialogue Modelling for Spoken Language Systems (Tutorial). In: ACL Workshop. (1997)

42. Allen, J.F., Byron, D.K., Dzikovska, M., Ferguson, G., Galescu, L., Stent, A.: Toward Conversational Human-Computer Interaction. AI Magazine 22(4) (2001)

43. Ferguson, G.M.: Knowledge Representation and Reasoning for Mixed-Initiative Planning. Technical Report TR562, University of Rochester (1995)

44. Smith, R.W.: Integration of Domain Problem Solving with Natural Language Dialog: The Missing Axiom Theory. In: Proceedings of Applications of AI X: Knowledge-Based Systems, Orlando, FL (1992) 270–278

45. Smith, R.W., Hipp, D., Biermann, A.W.: A Dialog Control Algorithm and its Performance. In: Proceedings of 3rd Conference on Applied Natural Language Processing, Trento (Italy) (1992) 9–16

46. Smith, R.W., Biermann, A.W.: An Architecture for Voice Dialog Systems Based on Prolog-Style Theorem-Proving. Computational Linguistics 21 (1995) 281–320

47. Spivey, J.M.: An Introduction to Logic Programming through Prolog. Prentice Hall (1996)

48. Ferguson, G., Allen, J., Biller, B.: TRAINS-95: Towards a Mixed-Initiative Planning Assistant. In: Proceedings of 3rd Conference on Artificial Intelligence Planning Systems (AIPS-96), Edinburgh (Scotland) (1996) 70–77

49. Allen, J.F.: The TRAINS Project: A Case Study in Building a Conversational Planning Agent. Journal of Experimental and Theoretical AI (JETAI) **7** (1995) 7–48

50. Heeman, P.A., Allen, J.F.: The TRAINS 93 Dialogues, TRAINS Technical Note 94-2 (1995)

51. Ferguson, G., Allen, J.F.: Arguing about Plans: Plan Representation and Reasoning for Mixed-Initiative Planning. In: Proceedings of 2nd International Conference on AI Planning Systems, Chicago, IL (1994) 43–48

52. Traum, D.R., Schubert, L.K., Poesio, M., Martin, N.G., Light, M.N., Hwang, C.H., Heeman, P.A., Ferguson, G.M., Allen, J.F.: Knowledge Representation in the TRAINS-93 Conversation System. Technical Report TN96-4, University of Rochester (1996)

53. Poesio, M., Ferguson, G., Heeman, P., Hwang, C.H., Traum, D.R., Allen, J.F., Martin, N., Schubert, L.K.: Knowledge Representation in the TRAINS System. Technical Report 663, University of Rochester (1994)

54. Schubert, L.K., Hwang, C.H.: Episodic Logic Meets Little Red Riding Hood: A Comprehensive, Natural Representation for Language Understanding. In: Natural Language Processing and Knowledge Representation: Language for Knowledge and Knowledge for Language. MIT/AAAI Press, Menlo Park, CA, and Cambridge, MA (2000) 111–174

55. Burstein, M., Ferguson, G.M., Allen, J.F.: Integrating Agent-Based Mixed-Initiative Control with an Existing Multi-Agent Planning System. In: Proceedings of the Fourth International Conference on MultiAgent Systems. (2000) 389–390

56. Ferguson, G.M., Allen, J.F.: Mixed-Initiative Dialogue Systems for Collaborative Problem Solving. In: Proceedings of AAAI Fall Symposium on Mixed-Initiative Problem Solving Assistants (FS-05-07), Washington, DC (2005) 57–62

57. Allen, J.F., Ferguson, G., Stent, A.: An Architecture for More Realistic Conversational Systems. In: Proceedings of 6th International Conference on Intelligent User Interfaces (IUI), Santa Fe, NM (2001) 1–8

58. Rich, C., Sidner, C.L.: COLLAGEN: When Agents Collaborate with People. In Johnson, W.L., Hayes-Roth, B., eds.: Proceedings of 1st International Conference on Autonomous Agents (Agents'97), New York, ACM Press (1997) 284–291

59. Rich, C., Sidner, C.L., Lesh, N.: COLLAGEN: Applying Collaborative Discourse Theory to Human-Computer Interaction. AI Magazine **22** (2001) 15–25

60. Rich, C., Sidner, C.L.: COLLAGEN: Middleware for Building Mixed-initiative Problem Solving Assistants. In Aha, G.D.W., Gecuci, G., eds.: Mixed-Initiative Problem Solving Assistants, Papers from the 2005 Fall Symposium, FS-05-07, Menlo Park, CA, AAAI Press (2005)

61. Grosz, B.J., Sidner, C.L.: Plans for Discourse. In Cohen, P., Morgan, J., Pollack, M.E., eds.: Intentions in Plans and Communication. MIT Press, Cambridge, Massachusetts (1990) 417–444

62. Litman, D.J., Allen, J.F.: A Plan Recognition Model for Subdialogues in Conversations. Cognitive Science **11** (1987) 163–200

63. Ferguson, G., Allen, J.F.: Generic Plan Recognition for Dialogue Systems. In: Proceedings of ARPA Workshop on Human Language Technology, Princeton, NJ (1993)

64. Rich, C., Sidner, C.L., Lesh, N., Garland, A., Booth, S., Chimani, M.: DiamondHelp: A Collaborative Interface Framework for Networked Home Appliances. In: Proceedings of 25th IEEE International Conference on Distributed Computing Systems Workshops, Columbus, OH (2005) 514–519

65. Wahlster, W., Reithinger, N., Blocher, A.: SmartKom: Multimodal Communication with a Life-Like Character. In: Proceedings of 7th European Conference on Speech Communication and Technology, Aalborg (Denmark) (2001) 1547–1550

66. Wahlster, W., ed.: SmartKom: Foundations of Multimodal Dialogue Systems. Springer (2001)

67. Bühler, D., Minker, W.: Mobile Multimodality - Design and Development of the SmartKom Companion. International Journal of Speech Technology 8(2) (2005) 193–202

68. Bos, J., Oka, T.: An Inference-Based Approach to Dialogue System Design. In: Proceedings of Conference on Computational Linguistics, COLING. (2002) 113–119

69. Blackburn, P., Bos, J.: Representation and Inference for Natural Language: A First Course in Computational Semantics. CSLI Publications (1999)

70. van Eijck, J., Kamp, H.: Representing Discourse in Context. In: Handbook of Logic and Linguistics. Elsevier (1996)

71. van Eijck, J.: Discourse Representation Theory. In: Encyclopedia of Language and Linguistics. Elsevier (2005)

72. Fitting, M.: First-Order Logic and Automatic Theorem Proving. Springer (1990)

73. Allen, J.F., Ferguson, G.: Actions and Events in Interval Temporal Logic. Technical Report TR521, University of Rochester (1994)

74. Ferguson, G.M.: Explicit Representation of Events, Actions and Plans for Assumption-Based Plan Reasoning. Technical Report 428, Computer Science Dept., University of Rochester (1992)

75. Fikes, R.E., Nilsson, N.J.: Strips: A New Approach to the Application for Theorem Proving to Problem Solving. In: Advance Papers of the Second International Joint Conference on Artificial Intelligence, Edinburgh (Scotland) (1971) 608–620

76. Barr, A., Feigenbaum, E.A.: The Handbook of Artificial Intelligence. Addison Wesley (1989)

77. Slaney, J.K.: FINDER: Finite Domain Enumerator - System Description. In: Proceedings of 12th International Conference on Automated Deduction (CADE-12), Nancy (France) (1994) 798–801

78. Bry, F., Yahya, A.: Minimal Model Generation with Positive Unit Hyper-Resolution Tableaux. In Miglioli, P., Moscato, U., Mundici, D., Ornaghi, M., eds.: Proceedings of Theorem Proving with Tableaux and Related Methods, 5th International Workshop, TABLEAUX'96, Terrasini, Palermo (Italy), Springer (1996)

79. Bry, F., Yahya, A.: Positive Unit Hyperresolution Tableaux and Their Application to Minimal Model Generation. Journal of Automated Reasoning 25 (2000) 35–82

80. Bry, F., Torge, S.: A Deduction Method Complete for Refutation and Finite Satisfiability. Lecture Notes in Computer Science 1489 (1998) 122–138

81. Gruber, T.R.: A Translation Approach to Portable Ontology Specifications. Knowledge Acquisition 5 (1993) 199–220

82. Baader, F., McGuinness, D.L., Nardi, D., Patel-Schneider, P.F., eds.: The Description Logic Handbook. Cambridge University Press, New York, NY (2007)

83. Kowalski, R.A., Sergot, M.J.: A Logic-Based Calculus of Events. New Generation Computing **4**(1) (1986) 67–95

84. Hayes, P.: A Catalog of Temporal Theories. Technical Report UIUC-BI-AI-96-01, The Beckman Institute, University of Illinois (1996)

85. Strauss, P.M., Hoffmann, H., Minker, W., Neumann, H., Palm, G., Scherer, S., Traue, H.C., Weidenbacher, U.: The PIT Corpus Of German Multi-Party Dialogues. In: Proceedings of 6th International Conference on Language Resources and Evalution (LREC), Marrakech (Morocco) (2008)

86. Kowalski, R.A.: The Early Years of Logic Programming. In: Communications of the ACM. Volume 31. (1988) 38–43

87. Bühler, D., Minker, W.: A Reasoning Component for Information-seeking and Planning Dialogues. The Kluwer International Series in Text, Speech and Language Technology. In: Spoken Multimodal Human-Computer Dialogue in Mobile Environments. Kluwer Academic Publishers, Dordrecht (The Netherlands) (2004)

88. Bühler, D.: Enhancing Existing Form-Based Dialogue Managers with Reasoning Capabilities. In: International Conference on Speech and Language Processing (ICSLP), Jeju Island (Korea) (2004) 3077–3080

89. Ramakrishnan, R., Ullman, J.D.: A Survey of Research in Deductive Database Systems. Logic Programming **2**(23) (1995) 125–149

90. Bühler, D., Hamerich, S.W.: Towards VoiceXML Compilation for Portable Embedded Applications in Ubiquitous Environments. In: Proceedings of European Conference on Speech Technology, EUROSPEECH, Lisbon (Portugal) (2005)

91. Bühler, D., Riegler, M.: Integrating Dialogue Management and Domain Reasoning. In: Proceedings of SPECOM. (2005) 409–412

92. Bühler, D., Hamerich, S.W.: Towards Embedding VoiceXML Applications through Compilation. In: Workshop Dialogsysteme mit XML-Technologien, Berliner XML Tage, Berlin (Germany) (2004)

93. Hamerich, S.W., Bühler, D.: Converging Dialogue Descriptions for Embedded and Server Platforms. In: Proceedings of 10th International Conference on Speech and Computer (SPECOM), Patras (Greece) (2005)

94. ECMA: ECMA-262: ECMAscript Language Specification. European Computer Manufacturers' Association (ECMA). (1999)

95. Mozilla.org: Rhino - JavaScript for Java. http://www.mozilla.org/rhino/ (1997)

96. Walker, W., Lamere, P., Kwok, P., Raj, B., Singh, R., Gouvea, E., Wolf, P., Woelfel, J.: Sphinx-4: A Flexible Open Source Framework for Speech Recognition. Sun Microsystems Technical Report (TR-2004-139) (2004)

97. Sun Microsystems Laboratories: FreeTTS. http://freetts.sourceforge.net (1999)

98. Bayer, S., Doran, C., George, B.: Dialogue Interaction with the DARPA Communicator Infrastructure: The Development of Useful Software. In: Proceedings of 1st International Conference on Human Language Technology Research (HLT 2001), San Diego, CA (2001)

99. Wahlster, W., ed.: Verbmobil: Foundations of Speech-to-Speech Translation. Artificial Intelligence. Springer (2000)

100. Reiter, E., Dale, R.: Building Natural Language Generation Systems. Cambridge University Press (2000)
101. Benzmüller, C., Horacek, H., Kruijff-Korbayová, I., Pinkal, M., Siekmann, J., Wolska, M.: Natural Language Dialog with a Tutor System for Mathematical Proofs. In: Cognitive Systems. Springer (2007)
102. : Owl 2 web ontology language document overview (2009)
103. Ranta, A., Cooper, R.: Dialogue systems as proof editors. In: IJCAR/ICoS-3. (2001) 225–240

Index

adversarial negotiation, *see* negotiation, adversarial

black-board
  communication, 149
  infrastructure, 149
  mechanism, 166
breadth-first search, *see* search, breadth-first

closed world assumption, 173
collaborative negotiation, *see* negotiation, collaborative
concurrent reasoning, *see* reasoning, concurrent
conflict
  detection, 21, 31, 131, 140
  handling, 121, 130, 140
  reformulation, 157
  resolution, 9, 10, 21, 84, 124, 130, 140, 141, 143, 161, 164, 168

deduction, 41, 44, 45
  algorithm, 84
depth-first search, *see* search, depth-first
dialogue
  analysis, 39
  control, 26, 30, 135
  design, 115, 119, 120, 124
  flow, 26, 119
  form-based, 23
  form-filling, 124
  frame-based, 151
  function, 8, 11

  history, 4
  initiative, 26
  interaction, 8, 9, 22, 25
  management, 4, 5, 8, 11, 12, 15, 17, 25, 28, 32–34, 38, 83, 84, 88, 102, 103, 105, 113–115, 117–119, 123, 124, 127–129, 131, 133, 138, 140, 144, 145, 147, 149–151, 153, 155, 156, 159, 161–164, 166–169, 172
  manager, *see* dialogue, management
  mixed-initiative, 136
  model, *see* dialogue, modelling
  modelling, 11, 12, 19, 24, 27, 36
  move, 4, 11, 12, 24, 25, 127, 128, 131–135, 143, 145, 147–149, 155, 156, 159, 164, 166
  participant, 16, 21, 24–26
  phenomenon, 9
  rule, 30
  script, 15, 22, 38, 119, 121
  scripting language, 11, 12, 15, 22, 127
  specification language, 24
  state, 22, 23, 84, 125, 127
  strategy, 27, 124
  structure, 20
  system, 1, 2, 4, 6, 8, 9, 12, 15, 17, 26–28, 34, 36, 37, 41, 55, 80, 83, 84, 115, 129, 148, 149, 161, 167, 168, 172, 173
  system-directed, 26
discourse, 3, 12, 15–21, 35–37, 161
  act, 19
  analysis, 16